Atmospheric Chemistry, Aging, and Dynamics

Atmospheric Chemistry, Aging, and Dynamics

Topic Editors

Zechen Yu
Myoseon Jang
Zhonghua Zheng

MDPI

Basel • Beijing • Wuhan • Barcelona • Belgrade • Novi Sad • Cluj • Manchester

Topic Editors

Zechen Yu
Hangzhou Institute for Advanced Study
University of Chinese Academy of Sciences
Hangzhou
China

Myoseon Jang
Department of Environmental Engineering Sciences
University of Florida
Gainesville
USA

Zhonghua Zheng
Department of Earth and Environmental Sciences
The University of Manchester
Manchester
UK

Editorial Office
MDPI AG
Grosspeteranlage 5
4052 Basel, Switzerland

This is a reprint of the Topic, published open access by the journals *Atmosphere* (ISSN 2073-4433), *Environments* (ISSN 2076-3298) and *Remote Sensing* (ISSN 2072-4292), freely accessible at: https://www.mdpi.com/topics/MM7L2CZ21F.

For citation purposes, cite each article independently as indicated on the article page online and as indicated below:

Lastname, A.A.; Lastname, B.B. Article Title. *Journal Name* **Year**, *Volume Number*, Page Range.

ISBN 978-3-7258-7747-8 (Hbk)
ISBN 978-3-7258-7748-5 (PDF)
https://doi.org/10.3390/books978-3-7258-7748-5

Contents

Wenjuan Li, Jian Wu, Yangyi Zhou, Qiongzhen Wang, Fuwei Yu and Rupei Wang
Chemical Characteristics and Sources Analysis of $PM_{2.5}$ in Shaoxing in Winter
Reprinted from: *Atmosphere* **2023**, *14*, 1256, https://doi.org/10.3390/atmos14081256 1

Giovanni Carabali, Faustino Juárez-Sánchez, Rafael N. Liñan-Abanto, Héctor Estévez, Mauro Valdés-Barrón, Roberto Bonifaz-Alfonso, et al.
Characteristics of Absorbing Aerosols in Mexico City: A Study of Morphology and Columnar Microphysical Properties
Reprinted from: *Atmosphere* **2024**, *15*, 108, https://doi.org/10.3390/atmos15010108 14

Fukang Qiao, Rongjie Zhang, Qiaojing Zhao, Fangfang Ma, Jingwen Chen and Hong-Bin Xie
A Surprisingly High Enhancing Potential of Nitric Acid in Sulfuric Acid–Methylamine Nucleation
Reprinted from: *Atmosphere* **2024**, *15*, 467, https://doi.org/10.3390/atmos15040467 31

Francesco D'Amico, Ivano Ammoscato, Daniel Gullì, Elenio Avolio, Teresa Lo Feudo, Mariafrancesca De Pino, et al.
Integrated Analysis of Methane Cycles and Trends at the WMO/GAW Station of Lamezia Terme (Calabria, Southern Italy)
Reprinted from: *Atmosphere* **2024**, *15*, 946, https://doi.org/10.3390/atmos15080946 45

Zhong Wang, Shengli Sun, Wenjun Xu, Rui Chen, Yijun Ma and Gaorui Liu
Research on Multiscale Atmospheric Chaos Based on Infrared Remote-Sensing and Reanalysis Data
Reprinted from: *Remote Sens.* **2024**, *16*, 3376, https://doi.org/10.3390/rs16183376 64

Yilong Huang, Jingwei Yang, Chang Liu, Ruiying Jing and Qiaohui Lu
Spatial and Temporal Variations in Rainwater Chemistry in a Rapid Urbanization Area of Shenzhen, China
Reprinted from: *Atmosphere* **2024**, *15*, 1536, https://doi.org/10.3390/atmos15121536 83

Lingli Fang, Yu Han, Junbo Lin and Wenkai Guo
Analysis of Forest Fire Emissions and Meteorological Impacts in Southwestern China Based on Multi-Source Satellite Observations
Reprinted from: *Atmosphere* **2025**, *16*, 1187, https://doi.org/10.3390/atmos16101187 99

Isabella Zaccardo, Antonella Buono, Teresa Laurita, Benedetto De Rosa, Francesco D'Amico, Davide Amodio, et al.
Can Saharan Dust Intrusions Alter the Isotopic Composition of Atmospheric Methane and Carbon Dioxide?
Reprinted from: *Environments* **2026**, *13*, 145, https://doi.org/10.3390/environments13030145 . . 120

atmosphere

MDPI

·icle

hemical Characteristics and Sources Analysis of $PM_{2.5}$ in haoxing in Winter

njuan Li [1,*], Jian Wu [2], Yangyi Zhou [3,4], Qiongzhen Wang [3,4,*], Fuwei Yu [3,4] and Rupei Wang [5]

1 College of Civil Engineering and Architecture, Zhejiang University of Water Resources and Electric Power, Hangzhou 310018, China
2 Ecological Environment Low Carbon Development Center of Zhejiang Province, Hangzhou 310012, China; wujian.ok@163.com
3 Eco-Environmental Science Research & Design Institute of Zhejiang Province, Hangzhou 310007, China; 13958100299@163.com (Y.Z.); fuwei.yu@hotmail.com (F.Y.)
4 Key Laboratory of Environmental Pollution Control Technology of Zhejiang Province, Hangzhou 310007, China
5 Zheneng Jinjiang Environment Holding Co., Ltd., Hangzhou 310011, China; wangrupei@hzjj.cn
* Correspondence: liwenjuan@zjweu.edu.cn (W.L.); wangqz03@126.com (Q.W.); Tel.: +86-0571-86929143 (W.L.); +86-0571-87997280 (Q.W.)

Abstract: By analyzing the mass concentrations and compositions of atmospheric $PM_{2.5}$ in Shaoxing from December 2019 to February 2020, the characteristics of carbon-containing components, water-soluble ions and metal elements were obtained. NO_3^-, OC, SO_4^{2-} and NH_4^+ were the main components of $PM_{2.5}$ in winter. The OC/EC ratio was 3.27, which proved the existence of SOC. The proportion of SOC in OC was 47.3%, which showed that secondary sources made a significant contribution. The values of OC/EC and NO_3^-/SO_4^{2-} indicated that vehicle exhaust emissions also made a significant contribution to $PM_{2.5}$. Trace elements of Na, Ca, K and Cd had higher enrichment factor values and were enriched due to human activities. Finally, $PM_{2.5}$ sources analysis was performed by the positive matrix factorization model. The results showed that secondary inorganic salts (49.3%), motor vehicles and industrial sources (21.3%) and dust sources (17.0%) were the important sources of $PM_{2.5}$ pollution.

Keywords: $PM_{2.5}$; metal elements; water-soluble ions; carbonaceous species; sources analysis

1. Introduction

At present, $PM_{2.5}$ has become the primary pollutant in most cities in China [1]. The primary aerosol particles and gaseous pollutants emitted by industry, transportation and fossil fuel combustion, as well as the secondary aerosols transformed by primary aerosols, are the main causes of $PM_{2.5}$ pollution [2]. The chemical components of $PM_{2.5}$ are complex, including water-soluble ions, organic carbon, elemental carbon and metal elements, mainly derived from human activities [3]. Water-soluble ions, especially secondary inorganic ions NH_4^+, SO_4^{2-} and NO_3^-, have become the main components of $PM_{2.5}$ in most cities in China [4].

In recent years, Zhejiang Province has successively conducted research on the components and sources of $PM_{2.5}$, mainly focusing on rapidly developing cities such as Hangzhou, Wenzhou and Ningbo. Li et al. focused on analyzing the pollution characteristics of carbon-containing components of $PM_{2.5}$ in winter in Hangzhou [5] and found that organic carbon accounted for a relatively high proportion. The compositional characteristics of water-soluble ions in $PM_{2.5}$ in Hangzhou were analyzed separately by Chen et al., and the results showed that SO_4^{2-}, NO_3^-, NH_4^+ and Cl^- were the main components [6]. Ge et al. analyzed the pollution characteristics of water-soluble ions in $PM_{2.5}$ in Wenzhou and found that SO_4^{2-}, NO_3^- and NH_4^+ were also the main components, mainly sourced from coal combustion, motor vehicle exhaust and biomass combustion [7]. Wang et al. conducted a study on the characteristics and sources of $PM_{2.5}$ pollution in Ningbo and found that the source categories that made significant contributions to $PM_{2.5}$ were secondary nitrates, secondary sulfates, marine sources, biomass combustion, high chlorine sources, heavy oil combustion, motor vehicle emissions, industrial smelting and mixed dust [8]. A targeted analysis on the spatiotemporal distribution of carbon-containing components in $PM_{2.5}$ was conducted by Du et al., also in Ningbo. The results showed that secondary organic carbon contributed significantly to the carbon components [9]. Xu et al. compared the pollution characteristics of polycyclic aromatic hydrocarbons in $PM_{2.5}$ in Hangzhou and Ningbo, and the results showed that the concentration of polycyclic aromatic hydrocarbons exceeded the national standard [10]. In addition, some scholars have begun to pay attention to the $PM_{2.5}$ pollution characteristics of small and medium-sized cities in Zhejiang Province. For example, Fang et al. studied the distribution characteristics of secondary organic aerosol indicators in $PM_{2.5}$ in Lanxi, which is a small city in Zhejiang Province [11]. Isoprene and toluene made significant contributions to secondary organic carbon (SOC).

Although extensive research has been conducted in many cities, there is still a lack of comprehensive studies on the chemical components and sources of $PM_{2.5}$ in Shaoxing. Shaoxing is a typical industrial city in Zhejiang Province. Zhang et al. detected the pollution characteristics of metal elements in $PM_{2.5}$ in Shaoxing [12]. Zhu et al. also analyzed the characteristics of organic carbon (OC) and elemental carbon (EC) since Shaoxing belongs to the urban agglomeration around Hangzhou Bay, which is the area with the most serious $PM_{2.5}$ pollution in the Yangtze River Delta [13]. The impact of anthropogenic sources on $PM_{2.5}$ pollution in Shaoxing is obvious.

Since 2013, due to the frequent occurrence of winter haze pollution, China has been committed to controlling particulate matter pollution, and the concentration of $PM_{2.5}$ has continued to decrease in recent years. However, according to the online monitoring data of air pollutants in China on http://www.aqistudy.cn/ (accessed on 26 July 2023), $PM_{2.5}$ pollution is still the most serious in winter. In winter, a $PM_{2.5}$ concentration exceeding the national standard limit often occurs in Shaoxing.

Therefore, this study focused on analyzing the chemical characteristics and sources of $PM_{2.5}$ in the atmosphere of Shaoxing from December 2019 to February 2020 by detecting the components such as carbon-containing components, water-soluble ions and metal elements. It would provide scientific support for the management of $PM_{2.5}$ pollution in Shaoxing during the winter.

2. Materials and Methods

2.1. Field Measurement

Considering urban functional area distribution, population density, environmental sensitivity and other factors, the environmental monitoring station of Shaoxing was selected as the sampling site. Shaoxing Environmental Monitoring Station is located at No. 38 Shuxiawang Road, which is a central area of commerce, transportation and residences.

Samples were collected from December 2019 to February 2020. Teflon filter membrane and quartz filter membrane were used for synchronous collection for 23 h. A 4-channel small-flow sampler and a medium-flow sampler were used to collect $PM_{2.5}$ samples, with sampling flow rates of 16.7 L/min and 100 L/min, respectively. The small-flow sampler collected four parallel samples (two quartz filter samples and two Teflon filter samples), and the medium-flow sampler synchronously collected one quartz filter sample, the brand of which was Waterman. A total of 25 groups of samples were collected.

2.2. Chemical Analysis

2.2.1. $PM_{2.5}$ Mass Concentration

Before sampling, a filter membrane was placed in a constant temperature and humidity chamber (temperature 20 °C and humidity 50%) for 24 h, and then an automatic weighing system (CR-4; Chinese Intelligent Manufacturing, Hangzhou, China) was used to weigh this filter membrane, and its mass was recorded. After sampling, the same instrument was used to weigh the same filter membrane and the mass was also recorded under the same conditions. The difference in mass was used to determine the mass concentration of $PM_{2.5}$.

2.2.2. Analysis of Water-Soluble Ions and Elemental Components

The concentrations of water-soluble ions were determined by ion chromatography. A 1/4 of one sample filter was taken and put into the sample bottle. The water-soluble components were extracted from the filter into 20.0 mL of deionized water, entered the sample bottle and soaked for 30 min. Then, the sample bottle was put into the ultrasonic instrument for ultrasonic extraction for 20 min. Ice was added into the ultrasonic instrument to ensure that the temperature was not higher than 20 °C. This could reduce the component loss. The extract was filtered by 0.45 μm microporous membrane filter and then sent to ion chromatograph (Ics-5000; Thermo Fisher, Waltham, USA) for analysis.

A filter sample collected by a small-flow sampler was put into a dry and clean sample box, and then analyzed by the WD-XRF wavelength dispersive X-ray fluorescence spectrometer (S4 pioneer; Bruker, Saarbrücken, Germany). Then, the concentrations of element components were determined.

2.2.3. Carbon-Containing Components Analysis

According to the Technical and Methodological Guidelines for Analytical Monitoring of Ambient Air Particulate Matter Sources (trial) (Second Edition) [14], the carbon-containing components were determined by the thermo-photometry method. A certain area of quartz sample filter membrane was put into a quartz boat and analyzed by the thermal optical carbon analyzer (DRI Model 2015; Desert Research Institute, Reno, USA). Firstly, under the condition of pure He, heated in temperature gradients of 140 °C (OC_1), 280 °C (OC_2), 480 °C (OC_3) and 580 °C (OC_4), all organic carbon in the sample was evaporated or decomposed, left the filter membrane and entered the oxidation furnace with the He gas flow (900 °C). The carbon element in the organic matter was oxidized to CO_2 by MnO_2. The CO_2 flowed out of the oxidation furnace with the He gas flow and was mixed with H_2. The mixture entered the reduction furnace (420 °C) and was reduced to CH_4 by Ni. Finally, the generated CH_4 was detected by flame ion detector (FID) to calculate the carbon content.

Then, He/O_2 mixed gas containing 10% O_2 was introduced, and the sample furnace was gradually heated up again. The sample was heated at 580 °C (EC_1), 740 °C (EC_2) and 840 °C (EC_3). In this process, the elemental carbon was oxidized in the oxidation furnace. The carbonaceous material was oxidized to CO_2, and then reduced to CH_4. Finally, the carbon-containing contents were calculated by detecting the generated CH_4 by FID. The total carbon (TC) mass concentration was the sum of the mass concentrations of OC and EC.

2.2.4. Quality Control and Quality Assurance

The testing process strictly followed the testing methods and procedures. Before injection, a solvent blank test and an experimental process blank test were conducted. The samples were injected in the order of $PM_{2.5}$ mass concentration, and no target compounds were detected in the blank test.

Recovery experiments were conducted during the detection of water-soluble ions. The standard solution was dropped onto the blank filter membrane. After the solution was air dried, the pretreatment process was performed following the sample operation process. Then, the blank filter membrane was tested on the machine. The added amount was equivalent to the actual concentration of the sample, and the results showed the recovery rates ranging from 80.0% to 120%.

2.3. Analysis Methods

2.3.1. Estimation of Enrichment Factor (EF)

The EF method was used to assess the man-made influence on metal elements. The calculation formula was shown as follows:

$$EF = \frac{(C_X/C_R)_{aerosol}}{(C_X/C_R)_{crust}} \tag{1}$$

where C_X was the mass concentration of element X, $\mu g/m^3$; C_R was the mass concentration of the reference element, $\mu g/m^3$. The subscripts aerosol and crust referred to the recipient sample and crust sample, respectively.

In this study, Al, which experiences less interference from human pollution, was used as the reference element [15]. When EF is less than 10, it indicates that these elements are not enriched and might be unaffected by human activities. When EF is larger than 10 and less than 100, it indicates that these elements are enriched to different degrees. When EF is larger than 100, it indicates that these elements are seriously enriched due to human activities [16].

2.3.2. Estimation of Sulfur Oxidation Rate (SOR) and Nitrogen Oxidation Rate (NOR)

SOR and NOR were used to characterize the conversion rates of SO_4^{2-} and NO_3^- in $PM_{2.5}$. The larger the SOR and NOR, the higher the secondary conversion efficiency [16]. The calculation equations were as follows:

$$SOR = \frac{C_{SO_4^{2-}}}{C_{SO_4^{2-}} + C_{SO_2}} \tag{2}$$

$$NOR = \frac{C_{NO_3^-}}{C_{NO_3^-} + C_{NO_2}} \tag{3}$$

where $C_{SO_4^{2-}}$ was the mass concentration of water-soluble sulfate ion, $\mu g/m^3$; $C_{NO_3^-}$ was the mass concentration of water-soluble nitrate ion, $\mu g/m^3$; C_{SO_2} was the mass concentration of SO_2 in the atmosphere, $\mu g/m^3$; C_{NO_2} was the mass concentration of NO_2 in the atmosphere, $\mu g/m^3$.

2.3.3. Calculation of Secondary Organic Carbon

The minimum OC/EC ratio was used to evaluate and verify the contribution of SOC to total organic carbon, and the specific calculation equation [17] was as follows:

$$SOC = OC - EC \times (OC/EC)_{min} \tag{4}$$

where $(OC/EC)_{min}$ was the minimum OC/EC value of the detection results of OC and EC.

3. Results and Discussion

3.1. $PM_{2.5}$ Mass Concentrations

During the sampling period, the average mass concentration of $PM_{2.5}$ in Shaoxing was 45.3 $\mu g/m^3$, which exceeded the national standard limit (35 $\mu g/m^3$, GB3095-2012) of 10.3 $\mu g/m^3$. The daily variation sequence of $PM_{2.5}$ concentrations manually sampled is shown in Figure 1. During the sampling period, there were a total of 15 days in which the daily average concentration of $PM_{2.5}$ exceeded the national standard limit, accounting for 60% of the total sampling days. The high daily average concentration ultimately led to exceeding the standard limit of the average concentration of $PM_{2.5}$ in winter.

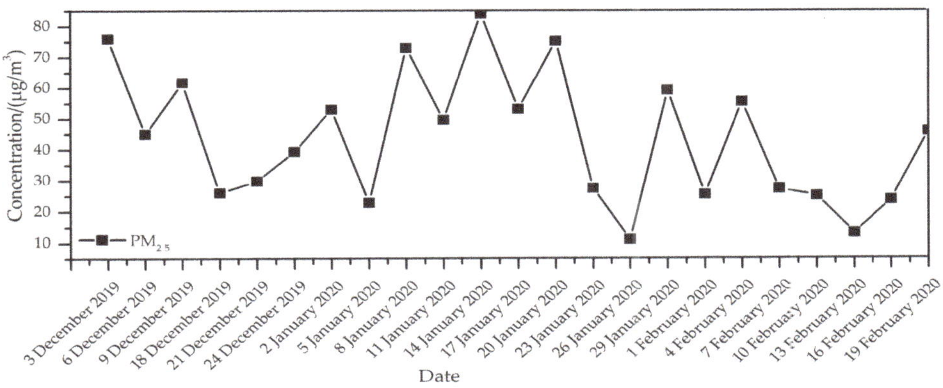

Figure 1. Daily variation sequence of $PM_{2.5}$ concentrations during the sampling period.

3.2. Carbon-Containing Components Characteristics

The daily variations in TC, OC and EC mass concentrations in winter in Shaoxing are shown in Figure 2. The average concentrations of TC, OC and EC were 9.86 μg/m^3, 7.41 μg/m^3 and 2.45 μg/m^3, accounting for 21.8%, 16.4% and 5.41% of $PM_{2.5}$ concentration, respectively. It could be seen that OC was the important component of $PM_{2.5}$.

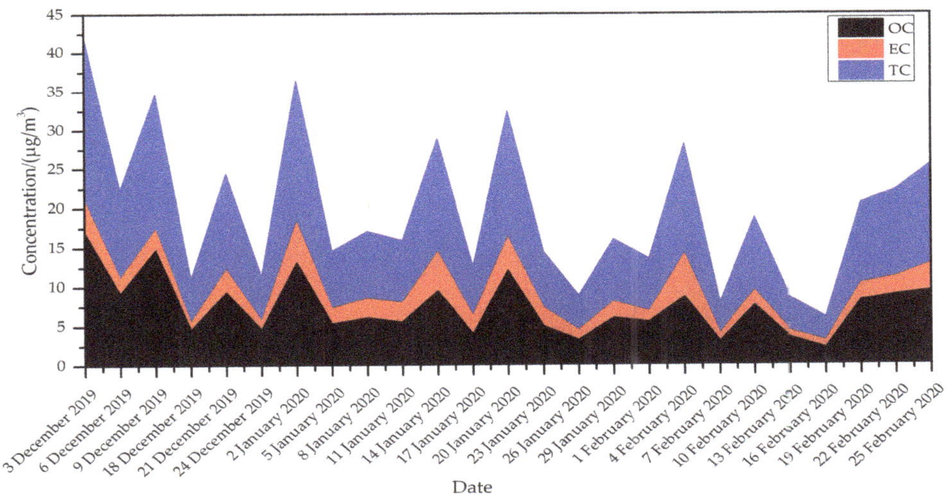

Figure 2. The daily variations in TC, OC and EC during the sampling period.

EC mainly comes from the incomplete combustion of fossil fuels or biomass, and only exists in the primary sources. OC sources include primary organic carbon (POC) emitted from coal, fuel and biomass combustion, and SOC formed by the conversion of VOCs and SVOCs [18]. That is to say, the POC sources in OC are consistent with EC sources. Therefore, the correlation analysis between OC and EC could be used to preliminarily determine the sources of carbon-containing components [19]. If the correlation coefficient between OC and EC is close to 1, it indicates that the OC and EC sources are consistent, both from primary sources [20]. If the correlation coefficient is less than 0.5, it indicates that the OC and EC sources are not consistent. As shown in Figure 3, the correlation coefficient R between OC and EC was only 0.71, indicating that the main sources of OC and EC in Shaoxing were not consistent. OC was more affected by SOC.

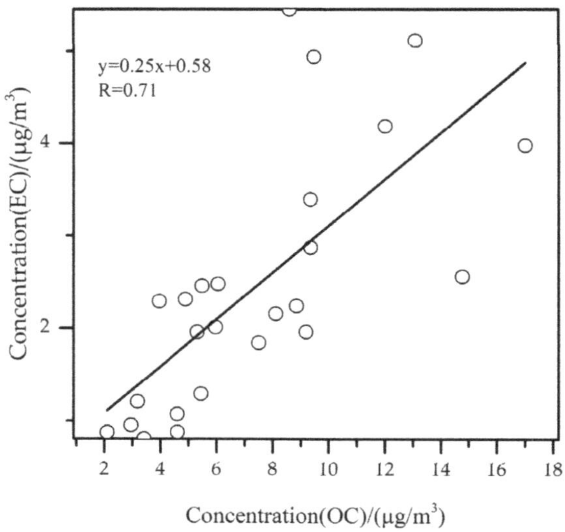

Figure 3. The correlation between OC and EC.

The ratio of OC/EC could reveal carbon-containing components' sources to some extent, and an OC/EC of 2.0 is often used as a basis for determining the presence of SOC [21]. The OC/EC ratio in this study was 3.27, indicating the presence of SOC in the carbon components of $PM_{2.5}$. The SOC was calculated using Equation (4), and the result is shown in Table 1. The average concentration of SOC was 3.54 $\mu g/m^3$, accounting for 47.8% of OC concentration and 7.82% of $PM_{2.5}$ concentration. This indicated that the contribution and impact of SOC were significant. Therefore, a more detailed analysis of the SOC characteristics should be conducted in the future, which will help to deepen the understanding of the $PM_{2.5}$ pollution characteristics in Shaoxing.

Table 1. The characteristics of SOC in winter in Shaoxing.

Season	OC ($\mu g/m^3$)	EC ($\mu g/m^3$)	OC/EC_{min}	SOC ($\mu g/m^3$)	SOC/OC
Winter	7.41	2.45	1.58	3.54	47.8%

Additionally, OC/EC is commonly used to preliminarily determine the types of primary and secondary sources of carbon-containing components. Through calculation, the OC/EC of Shaoxing was 3.27. According to the literature, when the OC/EC ratio is between 1.0 and 4.2, it indicates that exhaust emissions from diesel and gasoline vehicles exist [22,23]; meanwhile, 2.5 to 10.5 indicates coal-fired emissions contribute to carbon components [24]. Therefore, diesel vehicles, gasoline vehicles and coal-fired emissions all contributed to the carbon-containing contents of $PM_{2.5}$ in Shaoxing.

3.3. *Variation Characteristics of Water-Soluble Ions*

The water-soluble ions detected in this study included SO_4^{2-}, NO_3^-, F^-, Cl^-, Na^+, NH_4^+, K^+, Mg^{2+}, Ca^{2+}, etc. The mass concentrations of these ions are shown in Figure 4. The average total concentration of water-soluble ions during the sampling period was 30.6 $\mu g/m^3$. The highest average concentrations of NO_3^-, SO_4^{2-} and NH_4^+ were 14.9, 7.23 and 6.58 $\mu g/m^3$, respectively.

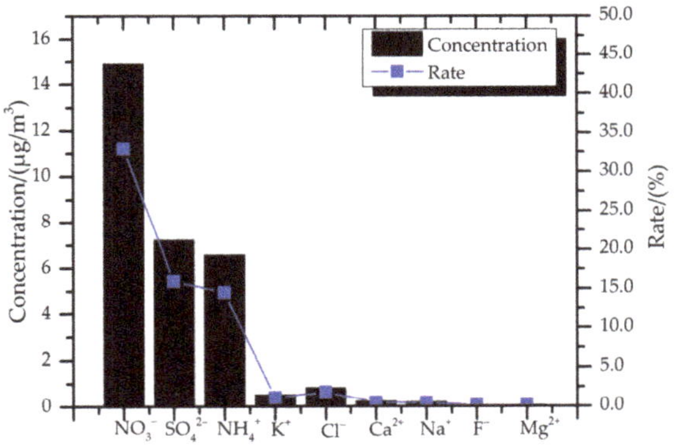

Figure 4. The average mass concentrations of water-soluble ions.

The total concentration of water-soluble ions accounted for 67.5% of $PM_{2.5}$ concentration. NO_3^-, SO_4^{2-} and NH_4^+ accounted for the highest proportions, accounting for 32.9%, 16.0% and 14.5%, respectively. This illustrated that NO_3^-, SO_4^{2-} and NH_4^+ had become the important components of $PM_{2.5}$ in Shaoxing. K^+ and Cl^-, as representative elements of coal-fired combustion, also had a place in $PM_{2.5}$ (a total of 2.94%). K^+ is also usually considered as a marker for biomass combustion. This proved that coal and biomass combustion made contributions to $PM_{2.5}$ in winter.

In Shaoxing, the proportion of NO_3^- was higher than that of SO_4^{2-}. This result was different from the reports about Hangzhou and Wenzhou [6,7]. The possible reason was that the emission sources had changed. The research about Wenzhou and Hangzhou was conducted in 2014 and 2015. According to the statistical yearbooks of these three cities, the numbers of vehicles in Hangzhou and Wenzhou in 2014 were 2.18 million and 1.57 million [25,26], respectively. Between 2014 and 2019, the number of vehicles in Shaoxing increased from 851,200 to 1.67 million [27,28]. Therefore, based on the large increase in the number of vehicles, the proportion of NO_3^- from vehicle exhaust in Shaoxing significantly increased.

NO_3^- and SO_4^{2-}, known as the secondary inorganic aerosols, are secondary reactions of primary pollutants such as SO_2 and NO_2 emitted into the atmosphere. In the past decade, due to the pollution problems caused by coal-fired sources (such as acid rain and particulate matter pollution), China has increased its efforts to govern SO_2 from coal-fired sources. The sources of NO_2 are more complex. The governance effect on NO_2 is not as good as SO_2. Therefore, the concentration of SO_2 has been maintained at a low level. According to the data from http://www.aqistudy.cn/ (accessed on 26 July 2023), the average concentration of SO_2 in the atmosphere of Shaoxing was only 7 $\mu g/m^3$ and the average concentration of NO_2 was 32 $\mu g/m^3$. This inevitably led to a significant increase in NO_3^-.

SOR and NOR reflect the conversion degrees of SO_2 and NO_2 into SO_4^{2-} and NO_3^- in the atmosphere, respectively. If NOR and SOR are larger than 0.1, it indicates that sulfates and nitrates are mainly produced by the photochemical oxidation of SO_2 and NO_2. Therefore, secondary pollution exists [29]. The higher the SOR and NOR, the higher the conversion efficiency. Thus, in this study, the SOR and NOR values were calculated using Equations (2) and (3), with average values of 0.61 and 0.34, respectively. This indicates that the conversion effect of SO_2 was stronger than that of NO_2.

Usually, with the concentration of particulate matter increasing, SO_2 is more likely to adsorb onto the surface of the particulate matter and undergo various homogeneous and heterogeneous reactions to generate secondary ions [30]. This might be the reason for the higher conversion rate of SO_2 in winter. There is a negative correlation between NOR and temperature [31], leading to a poor conversion of NO_2 under low temperature conditions in winter.

The correlations between various inorganic water-soluble ions in $PM_{2.5}$ can reflect the similarity in properties and sources between each ion [32]. NH_4^+ and NO_3^-, NH_4^+ and SO_4^{2-} had high correlations and the R values were 0.98 and 0.86, respectively (Figure 5). It indicates that these three ions mainly existed in the forms of $(NH_4)_2SO_4$ and NH_4NO_3 [32,33].

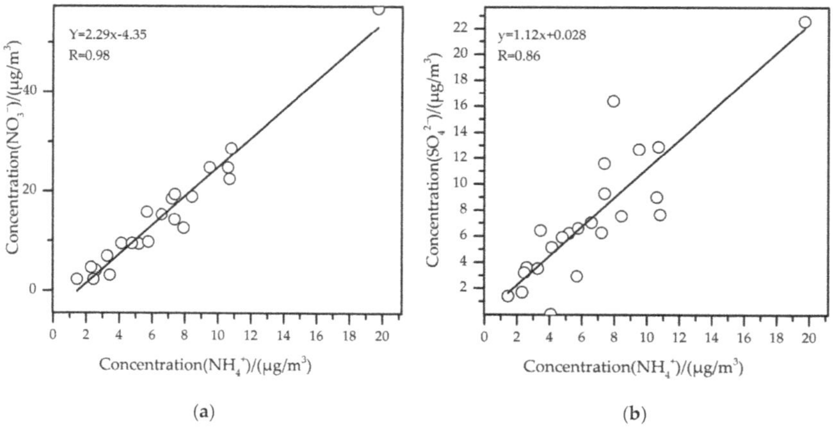

Figure 5. The correlations between NH_4^+ and NO_3^- (**a**), NH_4^+ and SO_4^{2-} (**b**).

The NO_3^-/SO_4^{2-} ratio could reflect the situation of vehicle exhaust and fixed combustion sources to some extent [31]. If the NO_3^-/SO_4^{2-} ratio is larger than 1, it indicates that the effect of vehicle exhaust is more obvious than that of fixed combustion sources; if the NO_3^-/SO_4^{2-} ratio is less than 1, it suggests that the contribution of fixed combustion sources is greater than that of vehicle exhaust [18]. The average ratio of NO_3^-/SO_4^{2-} was 2.06, suggesting that there were more sources to $PM_{2.5}$ emissions from vehicle exhaust compared to fixed combustion sources in the winter in Shaoxing.

3.4. Characteristics of Metal Elements

The mass concentrations and proportions of metal elements in $PM_{2.5}$ are shown in Table 2. K had the highest proportion, followed by Fe and Si. The background values of soil elements in Zhejiang Province [34] were selected to calculate the enrichment factors of metal elements, as shown in Figure 6. The EF values of Na, Ca, K and Cd were greater than 10, indicating that the impact of human activities was obvious and enriched these elements.

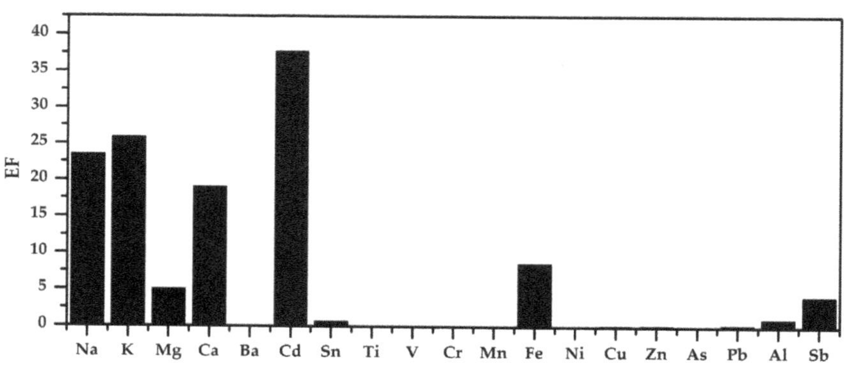

Figure 6. The EF values of metal elements.

Table 2. The mass concentrations and proportions of metal elements in $PM_{2.5}$.

Metal Element	Mass Concentration ($\mu g/m^3$)	Proportion in Metal Element (%)	Proportion in $PM_{2.5}$ (%)
K	0.514	10.72%	1.14%
Fe	0.300	6.26%	0.66%
Si	0.284	5.92%	0.63%
Na	0.159	3.31%	0.35%
Ca	0.152	3.16%	0.33%
Al	0.109	2.27%	0.24%
Zn	0.084	1.76%	0.19%
Pb	0.058	1.21%	0.13%
Cd	0.055	1.15%	0.12%
Mg	0.045	0.95%	0.10%
Ba	0.039	0.81%	0.09%
Mn	0.030	0.62%	0.07%
Sn	0.027	0.56%	0.06%
Sb	0.024	0.50%	0.05%
Cr	0.016	0.34%	0.04%
Ti	0.015	0.31%	0.03%
Cu	0.011	0.22%	0.02%

The sources of K were consistent with K^+ and might come from coal or biomass combustion. Coal or biomass combustion belong to anthropogenic sources. Na, Fe, Al and Si were crustal elements, mainly derived from crustal sources such as ground dust and soil fly ash. Although most of Fe came from crustal sources, it might be also influenced by human activities, such as steel smelting. In this study, the EF value of Na was greater than 10, indicating that crustal sources have caused Na to be enriched. Similarly, Ca was an indicator element for building construction [35]. The construction dust in Shaoxing has also enriched Ca. The EF value of Fe was less than 10, suggesting the influence of anthropogenic sources was weak. It indicates that Fe in $PM_{2.5}$ in Shaoxing was mainly from crustal sources.

Except for the above elements, the EF values of the other elements were all less than 10, indicating that they were not enriched.

3.5. $PM_{2.5}$ Sources Analysis

The positive matrix factorization (PMF) model was used to analyze the $PM_{2.5}$ sources in winter in Shaoxing. Based on the stability of the PMF analysis results, seven factors were identified as the optimal components spectra for various compounds in $PM_{2.5}$ (Figure 7).

The main contributions of factor 1 were Mg^{2+}, Ca^{2+}, Na^+ and $SO_4{}^{2-}$, which were inferred as the dusts from construction sites and roads. The main contributors to factor 2 were $NO_3{}^-$ and $NH_4{}^+$, which were secondary nitrates. The main contributor to factor 3 was L-glycan, which was inferred as biomass combustion [36]. Factor 4 was characterized by Cl^-, which was believed to mainly come from coal combustion in addition to natural sources. Therefore, factor 4 was inferred as the coal-fired source. Factor 5 was characterized by OC and EC, which were related to motor exhaust emissions. It also enriched metal elements such as Mn, Zn, Fe and Cu, which were related to industrial production activities [37]. Therefore, factor 5 was inferred as industrial sources and automotive exhaust. Factor 6 was characterized by Ti, Al and Si, which was inferred as soil dust. In this study, the dusts included road dust and construction dust, which belonged to man-made dust. The main contributors to factor 7 were $NH_4{}^+$ and $SO_4{}^{2-}$, which were secondary sulfates.

Among them, factors 1 and 6 were both related to the dusts, and factors 2 and 7 represented secondary inorganic salts. Thus, these factors were merged. As a result, the sources of $PM_{2.5}$ in Shaoxing were simplified into five categories: coal combustion, dust, biomass combustion, motor vehicle exhaust and industrial sources and secondary inorganic salts.

Figure 8 shows secondary inorganic salts had the highest proportion to $PM_{2.5}$ in winter in Shaoxing, with a contribution value of 49.3%. The proportions of motor vehicles and industrial sources and dust sources were 21.3% and 17.0%, respectively. Biomass burning also made a contribution, accounting for 7.79%, perhaps because of the biomass burning in autumn and winter in Shaoxing.

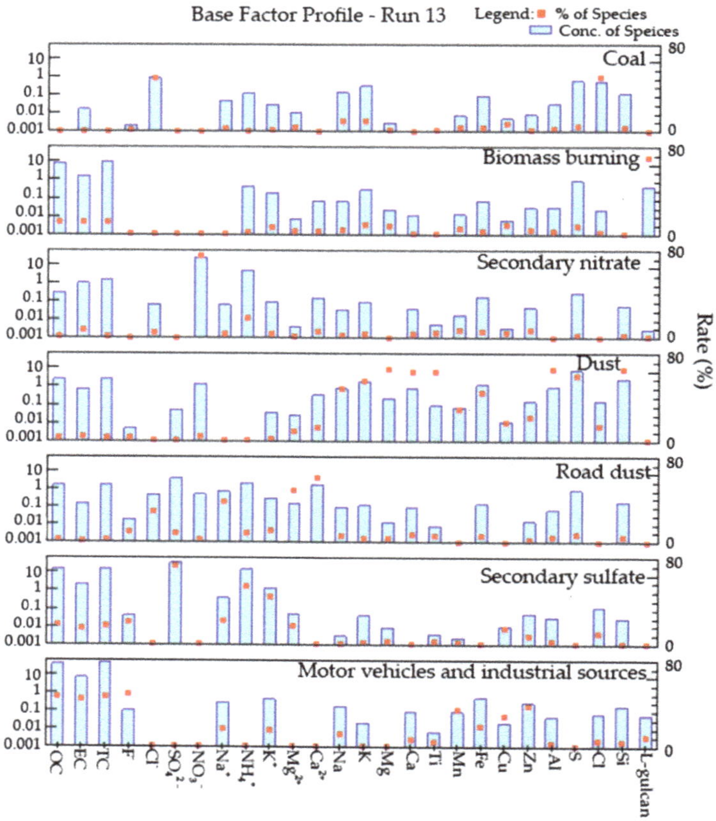

Figure 7. Factor profiles (% of species) of each source for the PMF model for $PM_{2.5}$.

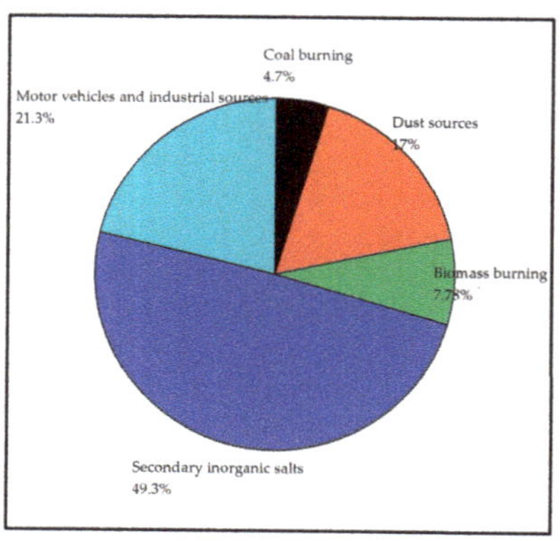

Figure 8. The result of sources analysis.

4. Conclusions

A total of 25 groups of $PM_{2.5}$ samples were collected to analyze the components in Shaoxing in winter. During the sampling period, the average concentration of $PM_{2.5}$ was 45.3 $\mu g/m^3$. There was an obvious pollution phenomenon. The chemical composition of $PM_{2.5}$ in winter was mainly composed of water-soluble ions and carbon-containing components.

Among all water-soluble ions, NO_3^-, SO_4^{2-} and NH_4^+ were three main components accounting for 63.4% totally of $PM_{2.5}$. These three ions mainly existed in the form of $(NH_4)_2SO_4$ and NH_4NO_3. Through the calculation of SOR and NOR, it could be found that a considerable portion of NO_2 and SO_2 in the air was converted into NO_3^- and SO_4^{2-}. The secondary conversion was very obvious. Based on the ratio of NO_3^-/SO_4^{2-}, it could be determined that the impact of vehicle exhaust on $PM_{2.5}$ pollution was more obvious than that of fixed combustion emission.

OC was the main component among the carbon-containing components. The OC/EC ratio was 3.27, indicating secondary aerosols existed in $PM_{2.5}$. After calculation, the average concentration of SOC was 3.54 $\mu g/m^3$, accounting for 47.8% of OC. The proportion of SOC in OC was relatively high.

The concentrations of metal elements in $PM_{2.5}$ were relatively low compared with carbon-containing components and water-soluble ions. However, the EF values of Na, Ca, K and Cd exceeded 10, indicating that these elements were obviously enriched due to human activities.

Finally, this study used the PMF model to analyze the sources of $PM_{2.5}$ in Shaoxing in winter. The result indicated that the $PM_{2.5}$ primary source was secondary inorganic salts (49.3%), followed by motor vehicles and industrial sources (21.3%) and dust sources (17.0%).

Author Contributions: Q.W. and F.Y. conducted the PMF analysis; J.W. conducted the sampling; W.L. prepared the original draft and analyzed the data; Y.Z. amended the language; R.W. completed the writing—review and editing. All authors have read and agreed to the published version of the manuscript.

Funding: This research was supported by Zhejiang Province Ecological Environment Research and Achievement Promotion Project (ZY0444202210018).

Institutional Review Board Statement: Not applicable.

Informed Consent Statement: Not applicable.

Data Availability Statement: All relevant data are within the paper.

Conflicts of Interest: The authors declare no conflict of interest.

References

Guo, X.B.; Wei, H.Y. Progress on the health effects of ambient $PM_{2.5}$ pollution. *Chin. Sci. Bull.* **2013**, *58*, 1171–1177. (In Chinese)

Yu, J.Y.; Wang, J.; Xu, L.P.; Li, L.; Liu, R.L.; Liu, J. Characteristics of chemical components of $PM_{2.5}$ and its seasonal variations in Chongqing urban area. *Chin. J. Environ. Eng.* **2017**, *11*, 6372–6378. (In Chinese)

Sun, Y.L.; Zhang, G.S.; Wang, Y.; Han, L.H.; Guo, J.H.; Dan, M.; Zhang, W.J.; Wang, Z.F.; Hao, Z.P. The air-borne particulate pollution in Beijing-concentration, composition, distribution and sources. *Atmos. Environ.* **2004**, *38*, 5991–6004. [CrossRef]

Li, X.R.; Wang, Y.S.; Guo, X.P.; Wang, Y.F. Seasonal variation and source apportionment of organic and inorganic compounds in $PM_{2.5}$ and PM_{10} particulates in Beijing, China. *J. Environ. Sci.* **2013**, *25*, 741–750. [CrossRef]

Li, L.W.; Dai, Q.L.; Bi, X.H.; Gao, J.X.; Yang, J.M.; Hong, S.M.; Feng, Y.C. Characteristics and sources of carbonaceous species in atmospheric $PM_{2.5}$ during winter in Hangzhou City. *Res. Environ. Sci.* **2017**, *30*, 340–348. (In Chinese)

Chen, J.Y.; Tang, K.J.; Zhu, Y.; Liu, B.C. Pollution characteristics of water-soluble ions in $PM_{2.5}$ of Hangzhou. *J. Zhejiang Univ. Technol.* **2016**, *44*, 410–416. (In Chinese)

Ge, L.L.; Zheng, Y.Z.; Tu, S.F.; Zhu, J.K.; Wang, Q.L.; Wang, X.Q.; Li, J.S.; Li, W. Characteristics and sources apportionment of water-soluble ions in $PM_{2.5}$ of Wenzhou, Zhejiang Province. *J. Zhejiang Univ.* **2017**, *44*, 112–120. (In Chinese)

Wang, W.F.; Ying, H.M.; Yu, J.; Zhou, J.; Xu, D.D.; Hu, M.; Du, B.H. $PM_{2.5}$ pollution characteristics and source analysis in Ningbo City. In Proceedings of the 9th Ningbo Academic Conference, Ningbo, China, 16 November 2016. (In Chinese).

9. Du, B.H.; Huang, X.F.; He, L.Y.; Hu, M.; Wang, C.; Ren, Y.C.; Ying, H.M.; Zhou, J.; Wang, W.F.; Xu, D.D. Seasonal and spa variations of carbon fractions in $PM_{2.5}$ in Ningbo and the estimation of secondary organic carbon. *Environ. Sci.* **2015**, *36*, 3128–3: (In Chinese)
10. Xu, H.H.; Xu, J.S.; He, J.; Pu, J.J.; Yu, K.A. Characteristics analyses of PAHs in $PM_{2.5}$ in the northern Zhejiang province. *Ch Environ. Sci.* **2018**, *38*, 3247–3253. (In Chinese)
11. Fang, Y.Z.; Zhou, M.T.; Jiang, M.; Huang, Y.N.; Feng, J.L. Characterization of secondary organic tracers in $PM_{2.5}$ in a representat city in mid-western Zhejiang Province. *Res. Environ. Sci.* **2018**, *31*, 2094–2102. (In Chinese)
12. Zhang, H.Y.; Xu, Q.E.; Yao, S.; Wang, Q.H.; Liu, S.J.; Li, H. Pollution characteristics and health risks of metal elements in ambi $PM_{2.5}$ in Shaoxing. *Mod. Chem. Res.* **2022**, *6*, 72–77. (In Chinese)
13. Zhu, J.; Wang, Q.Z.; Ding, H.; Li, H.; Yu, F.W.; Ji, Z.Q.; Jiang, Q.Q.; Wu, J.; Li, W.J. Characteristics and sources of organic carbon a elemental carbon in $PM_{2.5}$ in autumn and winter of Hangzhou Bay Area. *Environ. Pollut. Control* **2021**, *43*, 864–870. (In Chine
14. China Ministry of Ecology and Environment. The Guideline for Source Analysis and Monitoring Techniques of Environmental Particles. 2020. Available online: https://www.mee.gov.cn/xxgk2018/xxgk/sthjbsh/202005/t20200514_779089.html (acces on 14 May 2020). (In Chinese)
15. Hsu, S.C.; Liu, S.C.; Tsai, F. High wintertime particulate matter pollution over an offshore island (Kinmen) off southeastern Chi An overview. *J. Geophys. Res.* **2010**, *115*, 17309–17325. [CrossRef]
16. Yan, G.X.; Zhang, P.Z.; Huang, H.Y.; Gao, Y.; Zhang, J.W.; Song, X.; Zhang, J.Y.; Li, H.G.; Gao, Z.G.; Jiang, J.S.; et al. Concentrat characteristics and source analysis of $PM_{2.5}$ during wintertime in Zhengzhou-Xinxiang. *Environ. Sci.* **2019**, *40*, 2027–2((In Chinese)
17. Li, M.Y.; Yang, M.; Wei, M.; Zhu, H.X.; Liu, H.F. Characteristics and sources apportionment of fine particulate matter in a typi coastal city during the heating period. *Environ. Sci.* **2020**, *41*, 1550–1560. (In Chinese)
18. Lin, Y.; Ye, Z.X.; Yang, H.J.; Zhang, J.; Yin, W.W.; Li, X.F. Pollution level and source apportionment of atmospheric particles PM in southwest suburb of Chengdu in spring. *Environ. Sci.* **2016**, *37*, 1629–1638. (In Chinese)
19. Wang, Y.J.; Dong, Y.P.; Feng, J.J.; Guan, J.J.; Zhao, W.; Li, H.J. Characteristics and influencing factors of carbonaceous aerosols $PM_{2.5}$ in Shanghai, China. *Environ. Sci.* **2010**, *31*, 1755–1761. (In Chinese)
20. Turpin, B.J.; Huntzicher, J.J. Identification of secondary organic aerosol episodes and quantitation of primary and second organic aerosol concentration during SCAQS. *Atmos. Environ.* **1995**, *29*, 3527–3544. [CrossRef]
21. Wang, P.; Zhao, Y.Z.; Ding, W.C.; Li, Z.Z.; Liu, S.X. Chemical characteristics and source apportionment of $PM_{2.5}$ at Jinniu Dist of Chengdu city in winter. *China Powder Sci. Technol.* **2020**, *26*, 14–21. (In Chinese)
22. Schauer, J.J.; Kleeman, M.J.; Cass, G.R.; Simoneit, B.R.T. Measurement of emissions from air pollution sources. 2. C1 through C organic compounds from medium duty diesel trucks. *Environ. Sci. Technol.* **1999**, *33*, 1578–1587. [CrossRef]
23. Schauer, J.J.; Kleeman, M.J.; Cass, G.R.; Simoneit, B.R.T. Measurement of emissions from air pollution sources. 5. C1-C32 orga compounds from gasoline powered motor vehicles. *Environ. Sci. Technol.* **2002**, *36*, 1169–1180. [CrossRef] [PubMed]
24. Chen, Y.J.; Zhi, G.R.; Feng, Y.L.; Fu, J.M.; Feng, J.L.; Sheng, G.Y.; Simoneit, B.R.T. Measurements of emission factors for prim carbonaceous particles from residential raw coal combustion in China. *Geophys. Res. Lett.* **2006**, *332*, 382–385. [CrossRef]
25. Wenzhou Municipal Bureau of Statistics. Wenzhou Statistical Yearbook 2014. 2019. Available online: http://wztjj.wenzhou.g cn/art/2019/9/12/art_1467318_38058877.html (accessed on 12 September 2019). (In Chinese)
26. Hangzhou Municipal Bureau of Statistics. Hangzhou Statistical Yearbook 2014. 2015. Available online: http://tjj.hangzhou.g cn/art/2015/11/1/art_1229453592_3819408.html (accessed on 1 November 2015). (In Chinese)
27. Shaoxing Municipal Bureau of Statistics. Shaoxing Statistical Yearbook 2015. 2015. Available online: http://tjj.sx.gov.cn/art/2(/12/31/art_1229362048_3587783.html (accessed on 31 December 2015). (In Chinese)
28. Shaoxing Municipal Bureau of Statistics. Shaoxing Statistical Yearbook 2019. 2019. Available online: http://tjj.sx.gov.cn/art/2(/12/31/art_1229362048_3587790.html (accessed on 31 December 2019). (In Chinese)
29. Han, T.T.; Liu, X.G.; Zhang, Y.H.; Yu, Q.; Zeng, L.M.; Hu, M.; Zhu, T. Role of secondary aerosols in haze formation in summer the megacity Beijing. *J. Environ. Sci.* **2015**, *31*, 51–60. [CrossRef]
30. Behera, S.N.; Sharma, M. Degradation of SO_2, NO_2 and NH_3 leading to formation of secondary inorganic aerosols: environmental chamber study. *Atmos. Environ.* **2011**, *45*, 4015–4024. [CrossRef]
31. Arimoto, R.; Duce, R.A.; Savoie, D.L.; Prospero, J.M.; Talbot, R.; Cullen, J.D.; Tomza, U.; Lewis, N.F.; Ray, B.J. Relationships amc aerosol constituents from Asia and the north pacific during PEM-West A. *J. Geophys. Res. Atmos.* **1996**, *101*, 2011–2023. [CrossR
32. Chen, X.Q.; Chen, J.S.; Wu, S.P.; Lin, C.C. *The Causes and Control Mechanisms of Atmospheric Compound Pollution in Urt Agglomerations on West Coast of the Taiwan Strait*; Science Press: Beijing, China, 2014; p. 64. (In Chinese)
33. Zhao, Y.N.; Wang, Y.S.; Wen, T.X.; Dai, G.H. Seasonal variation of water-soluble ions in $PM_{2.5}$ at Changbai Mountain. *Environ.* **2014**, *35*, 9–14. (In Chinese)
34. Fan, Y.H.; Wang, Y.Q. Background characteristics of soil elements in four plains of Zhejiang Province. *Geophys. Geochem. Exp* **2009**, *33*, 132–134. (In Chinese)
35. Yang, F.M.; He, K.B.; Ma, Y.L.; Cadle, S.H.; Chan, T.; Mulawa, P.A. Characteristics of Mineral Component in Ambient $PM_{2.5}$ Beijing. *Environ. Sci.* **2004**, *25*, 26–30. (In Chinese)

Urban, R.C.; Alves, C.A.; Allen, A.G.; Cardoso, A.A.; Queiroz, M.E.C.; Campos, M.L.A.M. Sugar markers in aerosol particles from an agro-industrial region in Brazil. *Atmos. Environ.* **2014**, *90*, 106–112. [CrossRef]
Chen, P.F.; Bi, X.H.; Zhang, J.Q.; Wu, J.H.; Feng, Y.C. Assessment of heavy metal pollution characteristics and human health risk of exposure to ambient $PM_{2.5}$ in Tianjin, China. *Particuology* **2015**, *20*, 104–109. [CrossRef]

atmosphere

Article

Characteristics of Absorbing Aerosols in Mexico City: A Study o Morphology and Columnar Microphysical Properties

Giovanni Carabali [1,*], Faustino Juárez-Sánchez [1], Rafael N. Liñan-Abanto [2], Héctor Estévez [1], Mauro Valdés-Barrón Roberto Bonifaz-Alfonso [1], David Riveros-Rosas [1] and Adriana González-Cabrera [1]

[1] Geophysics Institute, National Autonomous University of Mexico (UNAM), Mexico City 04510, Mexico; tino@igeofisica.unam.mx (F.J.-S.); hrestevez@igeofisica.unam.mx (H.E.); mauro@igeofisica.unam.mx (M.V.-B.); bonifaz@unam.mx (R.B.-A.); driveros@igeofisica.unam.mx (D.R.-R.); gonzalezc@igeofisica.unam.mx (A.G.-C.)

[2] Departamento de Física, Facultad de Ciencias, Universidad Nacional Jorge Basadre Grohmann, Tacna 23003, Peru; rnlinan@unjbg.edu.pe

* Correspondence: carabali@igeofisica.unam.mx

Abstract: This paper presents an analysis of the morphology and columnar microphysical properties of atmospheric aerosols in Mexico City (MC) for the period 2022–2023. The morphological study focused on the structure description of soot particles and tar balls (TB). By transmission electron microscope (TEM) and scanning electrode microscope (SEM), voluminous soot aggregates mixed with TBs were observed. The chemistry shows that both soot and TBs are mostly carbonaceous species with well-defined morphologies. On the other hand, the columnar aerosol microphysical properties recovered from AERONET show that the particles have a bimodal aerosol size distribution (ASD) with two modes: fine and coarse. The ASD remains constant without showing significant seasonal changes, only with some variability for coarse particles. The aerosol optical depth (AOD) value is significantly high, typical of urban areas. The real (n) and imaginary (k) parts of the complex refractive index (CRI) were obtained from the photometric measurements. The CRI values show seasonal variations, with spring being the season with the highest values for n, while the highest values for k were measured in winter.

Keywords: atmospheric aerosol; aerosol microphysics properties; sun photometric observations; aerosol morphology; complex refractive index; absorbing atmospheric particles

1. Introduction

Atmospheric aerosols are a combination of solid or liquid particles that stay suspended in the air. These particles play a crucial role in the planet's radiation balance [1]. They scatter and absorb solar and infrared radiation in the atmosphere, leading to direct radiative forcing [1,2]. Aerosols also indirectly influence climate because they act as cloud condensation nuclei (CCN), which affect the microphysical properties of clouds [1,3]. To understand the complex interactions between aerosols, radiation, and clouds, as well as their impact on air quality, it is necessary to study their direct and indirect effects. The most effective way to understand these effects is through a detailed analysis of aerosol optical properties (AOP), which are largely dependent on key microphysical parameters, such as particle morphology, size distribution, and the complex refractive index [4,5].

The study of aerosol morphology provides valuable information about the particle structure, mixing state, and aging process [6–9]. A detailed description of the particle shape reveals information about its formation process and surface characteristics, which describes the ability to react chemically with other aerosols [8,10]. Numerous theoretical approaches have been developed to better understand the morphology of atmospheric particles. The calculation of dynamic shape factors (χ) is a frequently employed method, but it has its limitations. Specifically, it is not sensitive enough to detect the finer details of particle shapes, and it is necessary to know the particle density [11]. The methods that are effective in studying aerosol morphology involve calculating the fractal dimension (DF), which is a mathematical concept used to describe the complexity and self-similarity of irregular or fragmented geometric shapes [6–9]. In recent decades, the use of sophisticated electron microscopy techniques and the rapid advancements in computer language and image recognition techniques have made DF a widely used parameter to describe the morphology of soot particles [8]. Carabali et al. (2016, 2012) have employed the DF parameter to study aerosol transport in the MC pollutant plume and to describe the morphological distributions of soot emitted by wood-burning stoves [12,13].

The complex refractive index (CRI) is another essential microphysical parameter of aerosols, which is used in radiative transfer calculations [14,15]. CRI is generally described as m = n + ik, where n is the real part, describing the scattering characteristics of particles, while ik is the imaginary part that relates the absorption properties of aerosols. CRI is highly dependent on aerosol composition and is an important input for models simulating aerosol—electromagnetic radiation interactions. Measuring CRI for aerosols is complicated because its value can vary depending on factors such as aerosol composition, size, and atmospheric conditions [14,15]. Therefore, measurements and estimation techniques must take these variations into account to provide accurate results. A combination of spectroscopic methods (e.g., ultraviolet violet (UV) and infrared spectroscopies (IR)) and modeling is usually used to estimate the CRI in atmospheric particles [4,5]. For the MC basin, Dubovik et al. (2002) reported a mean CRI of 1.47-0.006i recovered from sun photometric measurements, while Moffet and Prather (2009) found a value of 1.49 for the n part of the CRI from theoretical Mie calculations [16,17]. These CRI values not only show a high load of these particles but also demonstrate the presence of highly scattering and absorbing aerosols in the MC atmosphere.

CRI values can be retrieved from spectral measurements of AOD, direct and diffuse radiation, using a sun photometer. This method has been successfully applied in the worldwide Robotic Aerosol Network (AERONET) [18,19]. The AERONET project, which was launched in the 1990s, is a global network that measures aerosol microphysical properties, such as aerosol size distribution, AOD, precipitable water vapor (PW), and the Angstrom exponent (AE). The primary goal of the project is to monitor AOP from the ground for the long term. Mexico City has been using a sun photometer associated with AERONET since 1999 to continuously measure AOP. This has contributed to a better understanding of air quality in the region [20].

This paper presents a study of the microphysical properties of atmospheric aerosols in MC. The morphological analysis of recently sampled particles evidences important structural details of soot and TB, the two most common types of particles in the atmosphere of the MC basin. The columnar optical properties of the aerosols were also studied. Specifically, the CRI and the ASD retrieved from the AERONET site in Mexico City were analyzed. This article is organized as follows: a brief description of the measurement site and instrumentation is presented in Section 2. The morphology and chemical characteristics of the individual particles are discussed in Section 3.1; the results obtained from the fractal distribution curves are described in Section 3.2; the optical parameters of the aerosols and their seasonal variability is presented in Section 3.3; the size distribution of the aerosols can be found in Section 3.4; a detailed description of the complex refractive index is given in Section 3.5; and in Section 3.6, an aerosol type classification based on the spectral measurements of AE and AOD retrieved from AERONET is discussed. Finally, Section 4 presents the main conclusions.

2. Materials and Methods

2.1. Study Region

The Mexico City (MC) Metropolitan Area, also called the MCMA, is a large region in central Mexico that includes MC and 60 other municipalities (Figure 1). With a population of over 20 million people, it is one of the largest and most densely populated metropolitan areas in the Americas [21]. This city is known for its poor air quality, which is caused by a combination of factors such as heavy traffic, industrial activities, and the city's complex geography. The observation equipment utilized in this research was placed on the roof of the Institute of Atmospheric Sciences and Climate Change (ICACC) building at the Universidad Nacional Autónoma de México (UNAM) campus, located in southern MC (19°20′ N, 99°10′ W, 2250 m. a. s. l.). MC experiences three distinct seasonal periods, spring, winter, and rainy, due to its subtropical climate. Spring is from March to May, followed by the rainy season from June to October, and finally, winter is from November to February. While the exact dates may vary, these three seasonal periods are widely acknowledged and accepted [22,23].

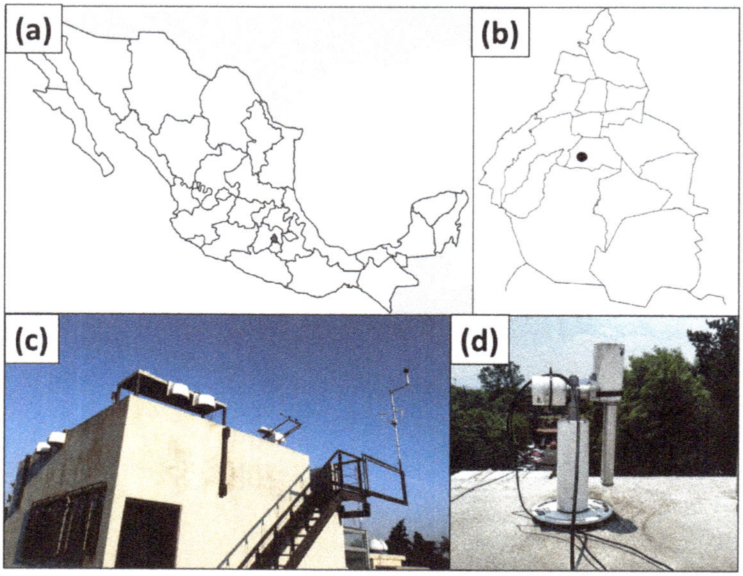

Figure 1. (**a**) Map of Mexico and geographic location of MC; (**b**) map of MC and location of SRO; (**c**) AERONET site in the SRO at UNAM main campus; and (**d**) image of the sun photometer CIMEL-318.

MC experiences three distinct seasons throughout the year due to its location in a subtropical zone. According to several meteorological studies, MC has three seasons: winter, from November to March; the spring season, from April to May; and the rainy season, from June to October [22,23]. The winter season is characterized by a dominant westerly current that brings an anticyclonic flow and clear skies, leading to a strong nocturnal thermal inversion [22,23]. This inversion persists until several hours after sunrise, when it is eroded by turbulent mixing generated by the strong solar heating of the surface. Spring is the warmest season due to higher sunshine, which leads to an increase in temperatures and evaporation rates, ultimately resulting in a drier environment. In April and May, the temperature usually reaches 30 °C with low relative humidity. Spring is also considered a dry season because rainfall tends to be lower compared to the summer months [22,23]. During the rainy season, prevailing easterly winds interact with the mountains surrounding Mexico City, forcing upward movement and convection. Clear mornings with infrequent thermal inversions mix pollutants with turbulent eddies. Convection over the mountains dilutes pollutants with vertical updrafts within the clouds. Precipitation has a large gradient within the basin, ranging from 400 mm to 1200 mm per year [22,23].

2.2. Aerosol Sampling

Aerosol particles were sampled in an 8-stage Micro Orifice Uniform Deposit Impactor (MOUDI, Model 100, MSP Corp.) [24]. MOUDI separates aerosol particles based on their aerodynamic size using eight impactor stages. Each stage consists of a set of nozzles and impactor plates that allow particles of a certain size range (from 0.8 to 10 μm) to be collected in each stage. Copper TEM grids with 200 mesh (Ted Pella Gilder grids, LF-200-Cu) coated with a thin film of collodion polymer were used to collect atmospheric particles. These grids were attached on 47 mm aluminum foils with double-sided carbon adhesive tape. Subsequently, the impact plates of the MOUDI were coated with aluminum foils equipped with the TEM grids. The MOUDI inlet flow rate was calibrated at 30 L/min, and sampling times of 20 min were set to avoid particle saturation on the TEM grids [12,25].

To study the morphology of atmospheric particles, aerosol samples were collected in the different seasons of the year: winter, spring, and rain. Sampling was performed in periods of three days for each season of both years, 2022 and 2023. The schedule of days sampled for each year is listed in Table 1. A total of 18 TEM grids with atmospheric particulate samples were obtained. Multiple aerosols are impacted within each TEM grid, but due to the operational limitations of the TEM, it was only possible to obtain 5 particle micrographs per TEM grid. In total, thirty particle images were obtained for each season, mainly winter and spring. It was not possible to obtain good particle images for the rainy season due to technical problems associated with the rupture of the polymer coating of the TEM grid and the humidity of some samples.

Table 1. Schedule of atmospheric aerosol sampling.

Season	2022	2023
Winter	January 18–20	January 24–26
Spring	May 03–05	May 7–9
Rainy	August 16–18	August 15–17

2.3. TEM and SEM Micrographs

Atmospheric particle micrographs for morphology studies were performed on SEM and HRTEM microscopes. These instruments are located at the Central Microscopy Laboratory (CML) of the Physics Institute (UNAM) [25]. SEM instruments (SEM JEOL-JSM-7800F) operate with voltages that fluctuate between 1 and 10 kV and are equipped with an energy-dispersive X-ray spectrometer (EDS) (Oxford Instruments, Abingdon, Oxfordshire, UK), which uses the AZtec 2.1 software for the analysis. The HRTEM instrument (JEOL JEM-2010F) operated at 200 kV near the Scherrer focus, with a theoretical point-to-point resolution of 0.20 nm and a spherical aberration of 0.5 mm. To capture images of the aerosols, TEM grids were directly introduced into the SEM and HRTEM without any prior treatment. The images were taken with a CCD camera and further processed using the GATAN digital micrographic system (version 3.7.0, Roper Technologies, Inc., Sarasota, FL, USA).

2.4. Fractal Dimension Analysis

Fractal dimension (DF) is a mathematical tool commonly used to describe the shape of non-spherical particles [9,26]. This concept was first introduced by the mathematician Benoit Mandelbrot, who assigned the term DF to Richardson's plot slope for the particle perimeter. The Richardson scheme or log—log plot is a graphical tool used to describe "self-similarity" and measure the perimeter of irregular shapes and objects [26]. The main characteristic of a fractal object is the self-similarity of its structure in different enlargements. This property can be studied by using the border-based fractal dimension equation [26,27].

$$L = K_L {\lambda_L}^{1-D_f} \tag{1}$$

where K_L is a constant, L is the perimeter, λ_L is the segment length (the length at which the perimeter is cut into small sections), and D_f is the fractal dimension.

The perimeter fractal dimension (PFD) of aerosol particles was calculated from the HRTEM micrographs. The calculation process begins with the processing of the digital particle images using the free software SCION Image 4.0, which optimizes the images and extracts the particle coordinates. SCION also provides the area of the particle and the length of its perimeter in pixels. Subsequently, these results were introduced into a computer code that calculate the fractal dimension of the particles [27]. In this study, a total of 60 particles collected in both the spring and winter seasons were analyzed. In each season, 30 particles were analyzed using a computational code that calculates the DF parameter of each particle. Based on the obtained DF results, fractal distribution curves (FDC) were plotted.

2.5. AERONET Station in Mexico City

The aerosol optical parameters were measured with a CIMEL sun photometer (CE-318), which is an automatic sun—sky-scanning spectral radiometer. The CE-318 was installed at the Solar Radiation Observatory (SRO), located on the roof of the ICACC building (Figure 1). Developed by NASA in 1992, the CIMEL is a state-of-the-art instrument designed to measure the optical properties of atmospheric aerosols. This sun photometer is a multichannel radiometer that measures the direct solar irradiance at various wavelengths (340, 380, 440, 500, 675, 870, 937, and 1020 nm). These channels cover a range of wavelengths from the ultraviolet (UV) through the visible and into the near-infrared (NIR) regions of the electromagnetic spectrum. Each channel is sensitive to different atmospheric components and aerosols, which allows for the characterization of aerosol properties and other atmospheric parameters such as the water vapor (PW) content. Numerous studies have previously described the instrumentation, data acquisition, retrieval algorithms, and calibration procedures in detail, which all conform to the rigorous standards of the AERONET global network [15,18,19,28].

The AOD values are calculated using the measurements of absorbed solar radiation in the spectral range of 340–1020 mn. This calculation is based on the Beer—Lambert—Bouguer law, which relates the attenuation of light to the concentration of absorbing and scattering particles in the atmosphere. Dubovik and King (2000) present an analysis of the uncertainty associated with the retrieval of AOD from AERONET [15]. The aerosol Ångström exponent ($\alpha_{440–870}$) (also referred to here as AE) is a parameter used to describe the spectral dependence of aerosol optical properties. It describes how the AOD changes with wavelength. The AE is calculated using measurements of aerosol optical depth at two or more wavelengths within the UV to NIR spectral range. In this work, precipitable water (PW), aerosol size distribution (ASD), and the complex refractive index (CRI) data with real (n) and imaginary (k) components were retrieved from the AERONET measurements for the studied period. For retrieving data from AERONET, there are three quality levels available: 1.0, 1.5, and 2. In this work, the level 1.5 database was utilized. This level contains raw measurements that are directly taken from the instrument, including direct sun/sky radiance and sky radiance measurements.

2.6. Aerosol Classification

The remote sensing formalism reported by Gobbi et al. (2007) was used to classify the aerosols and to know the predominant particle type in the MC atmosphere. In this method, the relationship between angstrom exponent ($AE_{(440, 870)}$) and its difference dAE ($AE_{(440, 675)}$—$AE_{(675, 870)}$) is plotted, which generates a scheme that allows for estimating the contribution of fine aerosol to the AOD [20,29].

3. Results

3.1. Morphology and Chemical Characteristics of Individual Particles

The study of atmospheric aerosols with high-resolution instrumentation can provide valuable insights into the particle physics characteristics that influence their optical properties and behavior in the atmosphere. Figure 2 shows the HRTEM images of aerosols that were sampled in MC's atmosphere at different periods of the year. Figure 2a shows the HRTEM image of a soot particle sample in the winter, which has a diameter (d_p) of 1.5 µm and is made up of chains of several nanometer-sized carbon particles that are interconnected. Likewise, when soot combines with larger particles, they can form more intricate structures such as agglomerates and fractal-like shapes. Figure 2b shows the HRTEM image of nanometer spherical particles (d_p < 700 nm) that are bound together. These types of particles are called tarball (TB) and are known for their unique spherical shape and composition, which mostly consists of carbon and oxygen. Nevertheless, depending on the emission source and atmospheric mixing process, they may also contain minor amounts of other elements [25]. According to a study conducted by Rios et al. (2023), the spring season in Mexico City is known for having a high number of forest fires, resulting in a large amount of TB in the atmosphere [25]. These particles are primarily made up of black or brown carbon and are mainly produced from biomass burning emissions. Something important that has recently been reported is the formation of voluminous carbonaceous aggregates, where the TBs are attached to the soot chains, as shown in Figure 2c [25]. Aside from their intricate morphology, these particles are also notable for their ability to absorb solar radiation efficiently. The EDS spectrum in Figure 2d reveals the elemental composition of the TB particles joined to soot aggregate (TB—soot aggregate in Figure 2c). According to the spectrum, these particles are mainly composed of carbon (C), followed by silicon (Si), and two comparatively weaker signals of oxygen (O) and sulfur (S). This composition is characteristic of both soot and TB particles.

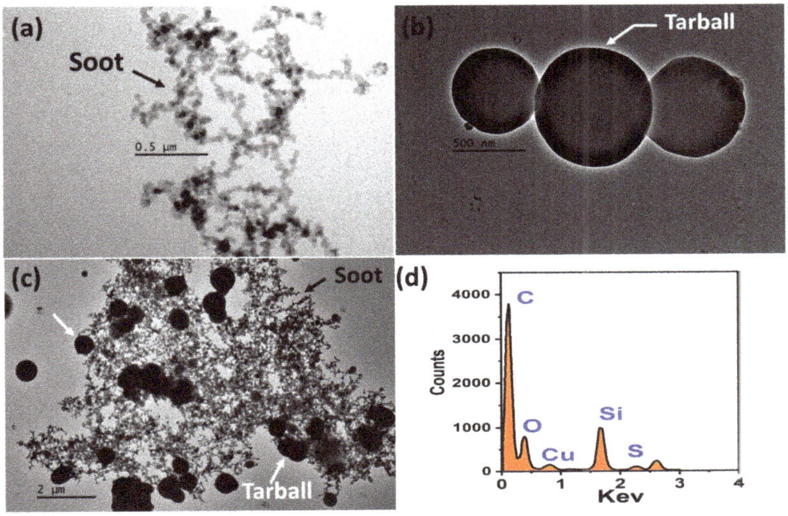

Figure 2. HRTEM images of (**a**) a soot particle, (**b**) three TB particles joined, and (**c**) a TB—soot aggregate. Panel (**d**) shows the EDS spectrum of the TB—soot aggregate.

Figure 3 shows the SEM micrograph of the TB—soot aggregate. These types of particles were very numerous on the TEM grids. The intricate arrangement observed in the soot—TB system could have been created via a coagulation process that occurred naturally in the atmosphere (Figure 3a). It is worth noting that inter-particle collisions within the MOUDI stage cannot be completely discarded during the sampling process. To prevent collisions and avoid saturation of the TEM grids, the sampling times were programmed for 15 min. Figure 3b also shows a magnification of the TB, where its almost spherical shape is observed. The elemental composition mapping reveals that the system is only composed of carbon, oxygen, and a low percentage of silicon (Figure 3c,d).

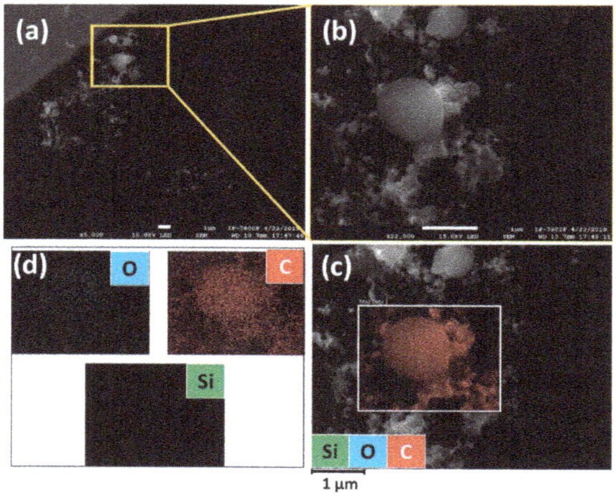

Figure 3. SEM image of (**a**) tarballs attached to soot aggregate and (**b**) tarball in high magnification mode. (**c**,**d**) correspond to the EDS elemental mapping.

3.2. Fractal Distribution Curves

Figure 4 shows the fractal distribution curves (FDC) of particles collected in the atmosphere of MC during spring and winter. According to the morphological model employed in this study, the fractal dimension values range from 1.00 to 1.40. Round-shaped particles tend to have DF values of around 1.00, while those with irregular shapes tend to have DF values that are close to 1.40. The FDC for the particles collected in the spring shows two modes: one at DF = 1.12 and the other at DF = 1.26. It seems that the first mode, with the high counts, corresponds to rounded-shaped particles, which might be associated with aged aerosols. On the other hand, the second mode (DF = 1.26), with a lower count value, represents particles with more complex shapes. These particles are most likely fresh emissions. The FDC for the particles collected during the winter showed two modes located at different values of the fractal scale. The first mode, at DF = 1.05, suggests a high presence of aged particles, while the second mode, around DF = 1.19—which is almost in the middle of the scale—reveals the presence of moderately irregular particles. These results suggest that the aging of aerosols is more likely to occur during winter due to favorable meteorological conditions. On the other hand, during spring, the FDCs observed a trend towards more irregular aerosol shapes, which could be attributed to the continuous pollutant emissions from forest fires. These emissions can lead to the formation of new particles that contribute to a high presence of soot and TB particles.

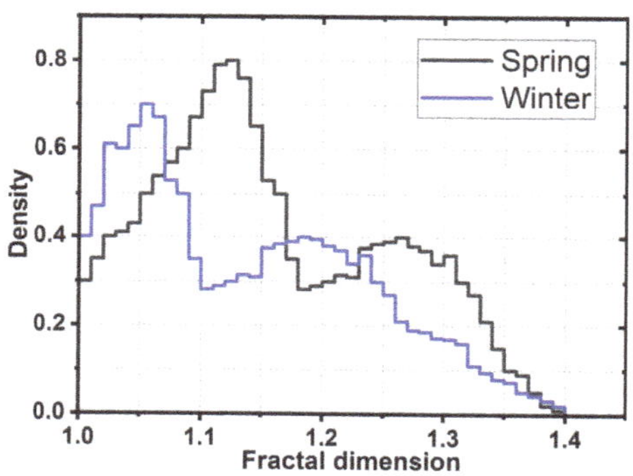

Figure 4. Fractal distribution curves (FDC) for atmospheric particles sampled in MC during the spring and winter seasons.

3.3. Aerosol Optical Parameters

This study analyzed the seasonal patterns of AE, PW, and AOD in MC during 2022 and 2023. The data were collected through 903 daily observations over two years. Figure 5 displays the time series of AE, PW, and AOD, while Figure 6 shows their monthly averages. The behavior of the AE, PW, and AOD parameters over the studied period can be observed through their respective time series. The AE, as shown in Figure 5a, is a parameter that characterizes the size distribution of atmospheric particles based on their spectral dependence on optical extinction. The AE value ranges from 0.8 to 1.9, with a mean value of 1.41. This variation in the AE value demonstrates the existence of both coarse and fine particles, the latter being the dominant type. Higher values of the AE (~1.49) measured during the spring months (March–May) indicate the dominance of fine fine-mode aerosols, which typically include particles like sulfates, nitrates, and organic matter, which are often found in higher concentrations in the atmosphere of the MC basin [30,31] The AE value (1.41) for winter was slightly lower than that obtained for the spring season. This result indicates that there is no substantial change in the particle size distribution between the winter and spring seasons despite the significant variation in weather conditions. The lowest AE value (1.37) was registered during the rainy season. The wet deposition processes during these months are highly favorable for capturing and removing aerosol particles from the atmosphere, which significantly impacts the AE value. The PW measurements in Figure 5b correspond to the total water vapor column in the atmosphere. The PW time series shows a cyclic variation, with high values registered between June and September. August has the highest PW values, with a maximum value of 1.7 cm. On the contrary, the spring and winter seasons present the lowest values of PW, with a mean of 1.0 cm.

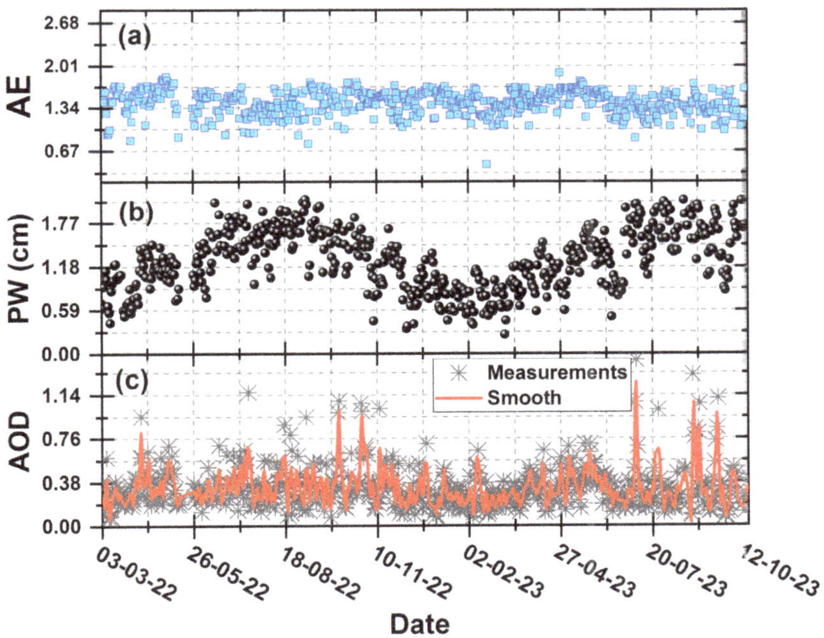

Figure 5. Time series of (**a**) precipitable water (PW), (**b**) Angstrom exponent (AE), and (**c**) aerosol optical depth (AOD) at 500 nm from direct Level 1.5 solar measurements conducted in the 2022–2023 period.

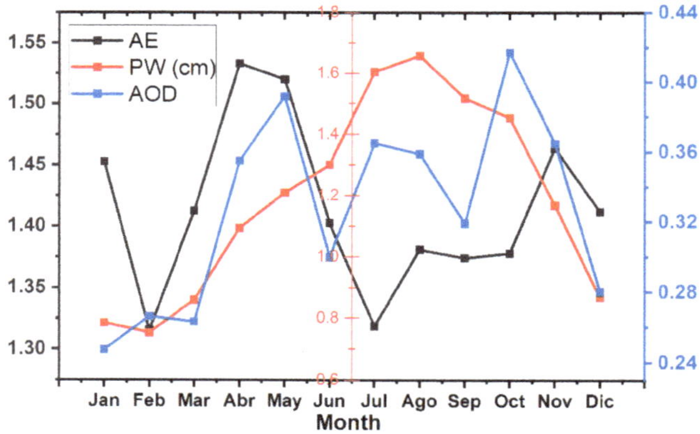

Figure 6. Monthly trend of AE, PW, and AOD (500 nm) parameters.

In the case of the AOD (Figure 5c), the results show that there is a significant monthly variation. October presented the highest mean AOD value of 0.42, followed by May with a mean of 0.39. However, it should be noted that the AOD values for the rainy months of July and August were 0.36 and 0.35, respectively. These values could have been affected by the presence of clouds during these months. It is important to note that photometric measurements can be susceptible to errors during the rainy season due to the presence of cloudy days. Carabali et al. (2017) indicate that the presence of cirrus clouds can affect the accuracy of AOD measurements, resulting in an overestimation during sun photometer measurements [20]. This phenomenon is known as "cirrus contamination" and has been extensively investigated [15,32]. The results obtained in the present work are comparable with those obtained by Carabali et al. (2017), who reported a climatological study of aerosol microphysical properties for the period 1999–2014.

3.4. Aerosol Size Distribution

The time series of the aerosol size distribution (ASD) obtained from sun photometric measurements at the MC AERONET site during the study period is presented in Figure 7. The ASDs were derived from daily measurements, and a total of 772 observations were considered for the analysis. The ASD provides important information about the aerosol particle sizes in the entire atmospheric column. By analyzing the shape of the distribution, it is possible to identify the types of aerosols present (fine and coarse particles), their size ranges, and how these distributions change over time. The colormap used to plot the distributions has "x" and "y" coordinates for Julian days and the size, while "z" indicates the volume of each mode (fine and coarse) in the distribution. The figures show that the ASD is bimodal, with one mode for fine particles with a mean size of around 0.18 μm and another for coarse particles of around 5.05 μm. The contour maps show that both fine and coarse particle modes are present throughout the year, with a significant variation in the volumetric concentrations that may be seasonally dependent. From Figure 8, it should be noted that although the ASD curves show a predominance of coarse particles, there are certain periods of the year, such as winter, when fine particles show a dominant presence (Figure 8).

In MC, coarse particles originate mainly from dust resuspension due to soil aridity and strong winds. These results contrast those reported by Carabali et al. (2017), where the predominance of fine particles is reported for the period 1999–2014 [20]. This slight reduction in the volume of fine aerosol in the atmospheric column could be attributed to the mitigation and control measures implemented by the environmental authorities in the past, as well as the partial lockdowns due to the pandemic situation at the beginning of 2022 [33,34].

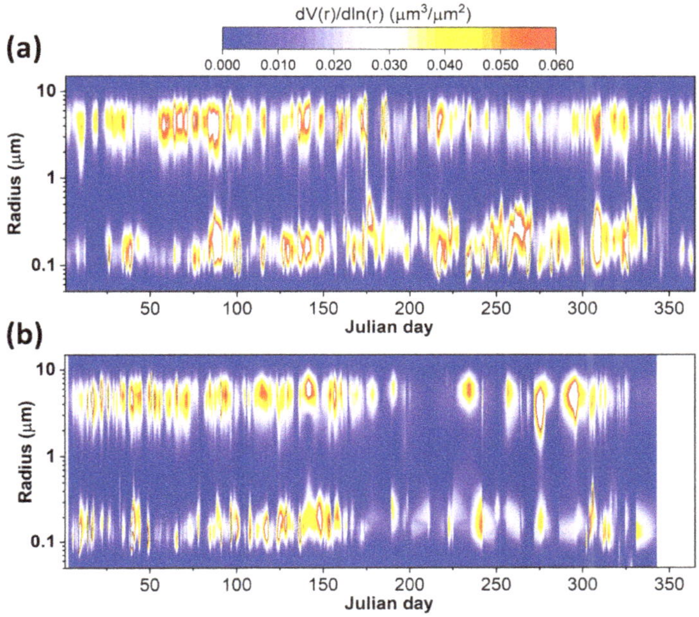

Figure 7. Aerosol size distribution (ASD) in the size range of 0.05–15.0 μm for (**a**) 2022 and (**b**) 2023.

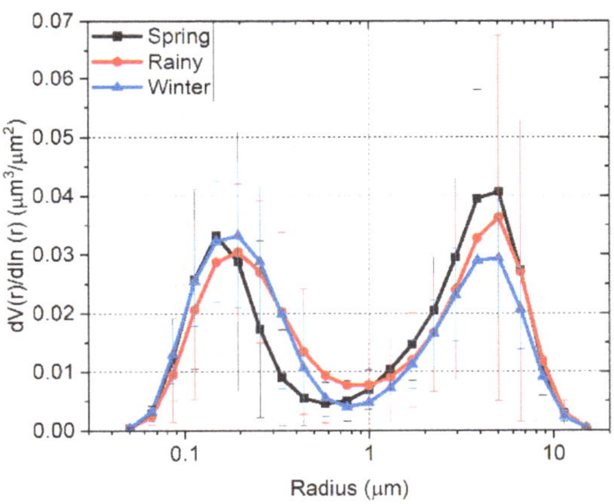

Figure 8. Seasonal ASD for the studied period.

Figure 8 shows the seasonal aerosol size distributions (ASD) for the studied period. The curves show a bimodal distribution where the coarse mode volume (V_c) predominates, except for the winter, where the fine particle volume (V_f) is slightly higher. The seasonal particle volumes in the fine mode are almost comparable, and only in the rainy season was the V_f slightly lower. Precipitation during the rainy season contributes to a decrease in the amount of particulate matter in the air. The coarse mode particle volume (V_c) shows significant seasonal variations. Spring is the season with the highest particle volume, followed by the rainy season. The lowest values of V_c were registered in winter. The high values during spring are due to soil resuspension and possibly to the formation and coagulation processes that favor an increase in aerosol size.

Figure 9 shows the average hourly time series of effective fine radius (R_{eff}) and effective coarse radius (R_{efc}) for the studied period. The time series trends show that R_{eff} remains stable throughout the year, maintaining a mean value of 0.16 μm. R_{efc}, on the other hand, shows considerable variability throughout the year, with a mean value of 2.90 μm. The constant radius value in the fine mode could be an indicator of the type of sources and their particulate emission dynamics. The urban area's high population density results in significant anthropogenic activity, which could explain the steady emission rate throughout the year. The R_{efc} values display a significant degree of variability, which is strongly influenced by meteorological changes. These R_{efc} fluctuations may be linked, to some extent, to the levels of precipitable water in the atmosphere. The results obtained are consistent with the findings of Saha and Moorthy (2004), who observed that changes in aerosol size are influenced by high relative humidity conditions [35].

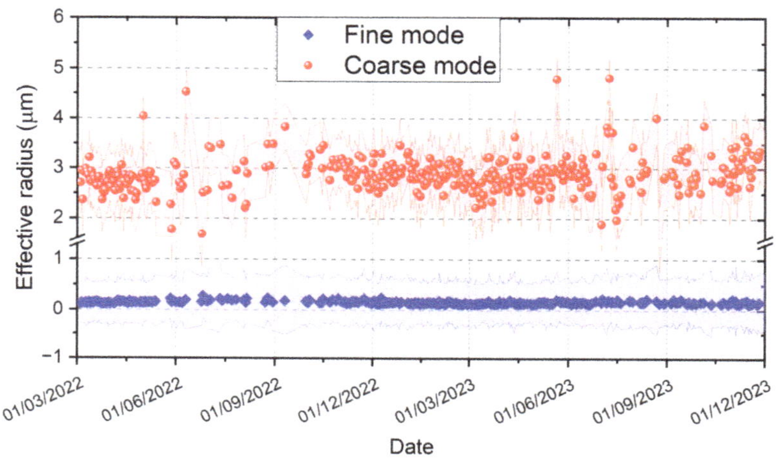

Figure 9. Average hourly time series of R_{eff} and R_{efc} for the studied period.

3.5. Complex Refractive Index

The optical properties of aerosols are influenced by their physical and chemical characteristics. Electromagnetic theory suggests that the CRI is a crucial physical parameter that determines how particles scatter and absorb radiation. The CRI is composed of two parts: real (n) and imaginary (k), which are values depending on the specific composition of the aerosols (e.g., dust, soot, water droplets, or various chemical compounds). Figure 10 presents the statistical analysis of the CRI measured during different seasons of the studied period. The box plots in the figure were constructed with daily observations of n and k parts of the CRI retrieved from the AERONET measurements. In addition, distribution curves are plotted next to the box plots, visually showing the dispersion of the data. The n values tend to exhibit a wider range of variations, while the k values are much more consistent. Figure 11 shows the wavelength dependence of the n and k parts of the CRI measured in the different seasons of the year. The data in Figure 11 show that the CRI has a spectral dependence, in which values for the n part tend to increase with the wavelength. It was observed that the rainy season had the lowest n values, while the highest was registered during the spring, followed by those values measured in the winter. On the other hand, the imaginary part of the complex refractive index shows a different trend, characterized by a stronger wavelength dependence, with minimum values measured at 675 nm. The highest values were obtained for the winter season, followed by those in spring. The lowest k values were measured in the rainy season, except for the values at 900 and 100 nm, which are slightly higher than those in winter.

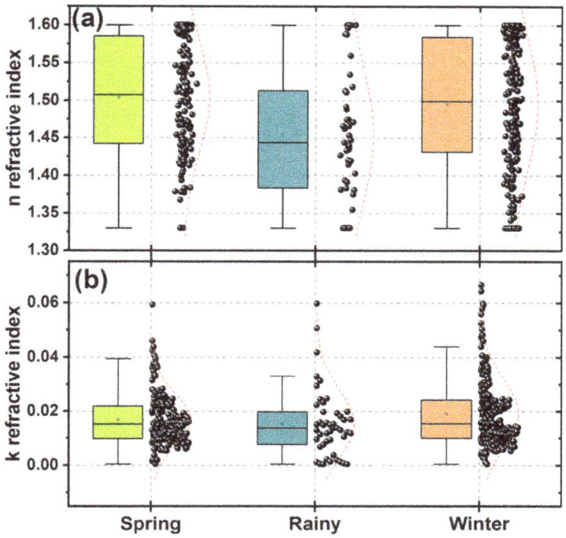

Figure 10. Statistics of seasonal values of the refractive index at 400 nm for (**a**) real (n) and (**b**) imaginary (k) parts. Box charts are plotted with black dots representing the individual measurements and red lines representing their normal distribution curve.

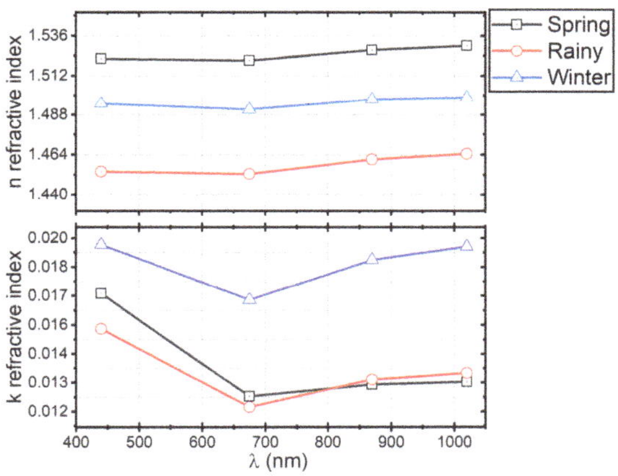

Figure 11. Wavelength dependence of the n and k parts of the CRI, measured in the different seasons of the year.

Table 2 shows the CRI values and other optical parameters of the atmospheric aerosols (AOD and AE) measured at different locations worldwide. For comparative purposes, the mean values found in this work are also listed. The values of seasonal n reported in this study are higher than those found in other urban areas, except for Paris (France), whose value (n~1.52) was measured in a different wavelength (532 nm), and Sahara (Africa), which has the highest value (n~1.53). Beijing and Kanpur show slightly lower n values than those observed for MC. One possible reason for this could be the variety of sources emitting within MC's urban area, which include mineral dust resuspended from soils by wind action, mainly in the spring [36]. In addition, there remains significant industrial activity within MC's air basin, which favors the emission of inorganic matter that enhances the scattering properties of aerosols [37]. The main reason behind the high value of n

that has been reported for the atmospheric particles in the Sahara is due to the dispersion characteristics of the prevailing type of aerosols in the region, whose particles are primarily mineral dust from the desert [38].

Table 2. Comparison between the values of the CRI and other aerosol optical parameters obtained in this study and those determined for other sites.

Site	*n*	*k*	AOD (550 nm)	AE	Date of Studies	Reference
Beijing, China	1.48	0.010–0.013	0.56–1.58	0.88–1.22	2005–2009, 2011–2012, 2014–2015	[14,39,40]
Kanpur, India	1.40–1.48	0.014–0.015	0.50–0.95	1.35–042	2011–2012	[39]
Shanghai, China	1.39–1.45	0.007–0.014	0.72	1.10–1.60	2010–2012	[41]
Sahara, Africa	1.53	0.0014	0.1–0.6	-	2006	[42,43]
Paris, France	1.51 (532 nm)	0.017 (532 nm)	0.15	-	2000, 2009–2010	[44,45]
Sao Paulo, Brazil	1.44	0.014	0.14–0.53	0.85–1.58	2007	[46]
Mexico City	1.49	0.017	0.33	1.41		This study

The results of this study show that the value of *k* is higher than those reported for the different cities (Table 2). The result obtained for MC is only comparable to that reported for Paris, although the measurement in that city was taken at 532 nm. Furthermore, the seasonal maximum value of 0.019 (440 nm) during winter in MC suggests that the *k*-value could potentially be influenced by weather conditions at that time of the year. During winter, the height of the boundary layer is reduced due to low temperatures, which leads to an increase in the concentration of aerosols in the vertical column of the atmosphere. These favorable conditions for aerosol mixing have a significant impact on the microphysical properties of the particles. In winter, for instance, the size distribution showed a bimodal curve dominated mainly by fine particles. Although, by definition, the CRI is an intrinsic property of the particles that do not depend directly on their concentration, its value can be influenced by the particle diameter, the degree of agglomeration of the particle, and its size distribution in the atmosphere. The high *k*-value found in this analysis is a clear indication of the strong absorbing nature of aerosols in MC, which poses a significant risk to the radiative balance of the urban area. In highly populated urban areas like the MC basin, the value of *k* is primarily attributed to the existence of soot and organic matter in the aerosol particles—as well as minerals and other compounds present in the aerosol, which contribute to a lesser extent.

3.6. Aerosol Classification: Identification of Major Aerosol Types

The graphical method proposed by Gobbi et al. (2007) was used to classify the aerosols measured in MC during the studied period [29]. The method visually examines the relation between AE and dAE to identify the fine aerosol contribution to the AOD, which, in turn, provides the respective particle size. This information can be very useful in various applications. For instance, it can be used to track mixtures of anthropogenic aerosol with dust, discern aerosol growth from cloud contamination, and observe aerosol humidification. Additionally, negative values of the difference (dAE) in the scheme indicate the predominance of fine-mode aerosols, while positive differences demonstrate a bimodal distribution of particles [29].

The data presented in Figure 12 are derived from a comprehensive analysis of 3500 direct sun photometer observations collected at the AERONET site in MC between 2022 and 2023. By using a color-coded system, the plot clearly shows the contribution of fine particles to AOD, with the blue dots indicating AOD values ranging from 0.15 to 0.3, green dots indicating AOD values ranging from 0.3 to 0.4, red dots indicating AOD values ranging from 0.4 to 0.7, and black dots indicating AOD values ranging from 0.7 to 1.0. Additionally, the plot presents continuous black lines depicting a fixed fine-mode size (Rf), while the dashed blue lines indicate the percentage contribution of particles to the AOD.

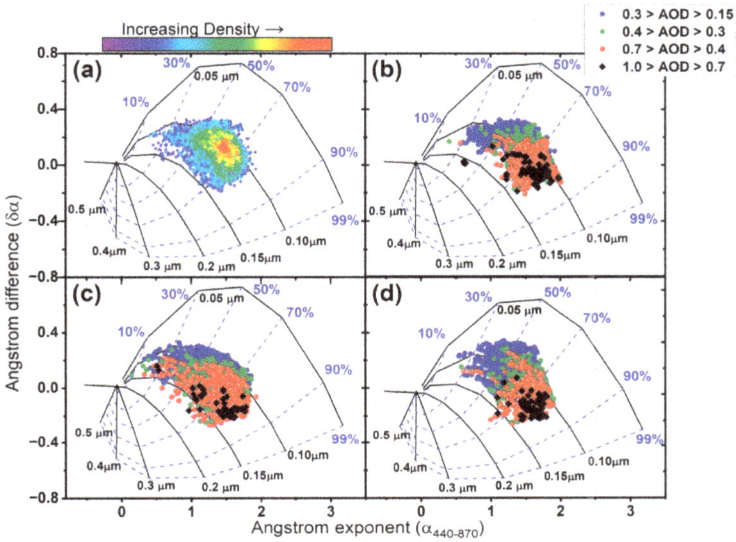

Figure 12. Relationship between dAE ($AE_{(440, 675)}$–$AE_{(675, 870)}$) and $AE_{(440, 870)}$ for MC aerosol measured during (**a**) the whole studied period, (**b**) spring, (**c**) rainy, and (**d**) winter seasons.

Figure 12a depicts all the data used in the analysis. It is interesting to note that the region with the highest density of data is situated at AE~1.5 with positive values of the difference at dAE~0.12. This observation suggests that the most significant contribution to the AOD in Mexico City (which is less than 70%) is due to fine particles with an approximately fixed radius (Rf) of 0.10 µm. The positive region of dAE indicates that the aerosol size distribution tends to be bimodal. Based on the schematic in Figure 1a, it can also be inferred that the coarse mode particles contribute to approximately 30% of the AOD.

Figure 12b–d represents the aerosol classification for the spring, rain, and winter seasons, respectively. The data shown in the diagrams are significantly different in terms of the number and distribution. However, it is important to note that there are several common features present in all of the diagrams. One of the most significant features is that almost 80% of the particles in each scheme fall within the range of radii between 0.10 and 0.15 µm. Based on the analysis, it can be inferred that there is no significant temporal variation in the mean particle size. This observation is in line with the finding for the R_{eff}, which had a nearly constant value of 0.16 throughout the analysis period. It is worth mentioning that this consistency in the mean particle size and R_{eff} values indicates a steady particle size distribution.

Another interesting fact is that in Mexico City's atmosphere, there are aerosols that generate AOD values of < 0.7 throughout the year, which is considered significantly high. These high extinctions indicate the absorbing nature of the aerosol and can be associated with hygroscopic growth and/or coagulation by aging of the aerosols in a fine mode [29]. However, it is important to note that some data with extinctions between 0.1 and 0.7 in the rainy season scheme are probably caused by cloud contamination effects. Carabali et al. (2017) found that there is a positive correlation between water vapor and high AOD values in the rainy season in MC [20].

4. Conclusions

The revelation of TB being attached to complex soot structures is one of the most interesting results of this work. These structures, which are referred to here as "TB-soot aggregate", consist of voluminous carbonaceous agglomerates with intricate structures. The amount and type of emission sources, such as old diesel-burning engines and biomass burning, which are present in the MC area, along with specific meteorological conditions within the MC basin, may create a favorable environment for the TB—soot aggregate formation in the atmosphere. Despite containing lower levels of other chemical elements, these particles are predominantly composed of carbon, which evidences its absorptive characteristics.

Upon calculating the particle fractal dimension, some slight seasonal variations in the morphological behavior of the particles were observed. These variations could be attributed to the influence of meteorological conditions during different seasons. Furthermore, it was noted that during the winter season, the particles tend to undergo a more pronounced aging process. This observation may be related to an increase in particle concentration due to the low mixing layer height. During the spring season, the number of aerosol particles with irregular shapes is notably higher. This increase can be attributed to the continuous emissions of aerosol particles caused by forest fires. Furthermore, the fractal distribution curve for spring indicates that there is a significant presence of aged aerosol particles. These aged particles are more likely to form under the influence of high solar radiation intensity, which is prevalent during the spring months.

In this study, columnar optical characteristics of aerosol in MC from 2022 to 2023 were analyzed using sun photometer observations. It found that the global mean AOD (500 nm) is 0.34, which is typically observed in urban areas. The mean AE value of approximately 1.41 indicates the presence of two particle modes (fine and coarse) that remain relatively constant throughout the analyzed period. The coarse mode dominates the size distribution, except during the winter season. Further analysis reveals that the specific radii of the fine and coarse modes are 0.16 and 2.90 μm, respectively, with the latter exhibiting significant variations throughout the year. The average CRI value for the period studied was 1.49-0.017i (400 nm). For MC, the CRI presented a seasonal variation, with the highest values in winter and spring. The low values observed in the rainy season are due to a reduction in contamination.

Finally, according to the microphysical observations of aerosol in Mexico City, it can be stated that the particles tend to be significantly absorbing. This can be evidenced by the LRC, whose value (1.49-0.017i) is higher than those measured in other urban areas. In addition, it was found that fine particles with radii in the range of 0.10–0.15 μm make the most significant contribution to the AOD in Mexico City, representing almost 80% of it.

Author Contributions: Conceptualization, G.C.; methodology, R.N.L.-A., H.E, M.V.-B. and R.B.-A.; software, D.R.-R. and A.G.-C.; validation, G.C., F.J.-S. and M.V.-B.; formal analysis, G.C.; investigation, G.C.; resources, H.E.; data curation, R.B.-A.; writing—original draft preparation, G.C.; writing—review and editing, G.C.; project administration, G.C.; funding acquisition, G.C. All authors have read and agreed to the published version of the manuscript.

Funding: This research was funded by UNAM-DGAPA PAPIIT, grant IN113322 (Mexico), and partial support was provided by Instituto de Geofísica (UNAM), grant R104 (internal projects).

Institutional Review Board Statement: Not applicable.

Informed Consent Statement: Not applicable.

Data Availability Statement: The aerosol microphysical parameters were retrieved from the AERONET site in MC. The data are available at the following website: https://aeronet.gsfc.nasa.gov/ (accessed on 20 July 2023). For information about the morphological analysis code, please contact the first author.

Acknowledgments: The authors thank B. Holben and the AERONET staff for their sun photometer calibration and support. Special thanks to the CML personnel at UNAM for their support during the operation of the TEM and SEM microscopes. Also, special thanks to the chemist Maria Isabel Saavedra for her support with the filters for the aerosol sampling.

Conflicts of Interest: The authors declare no conflicts of interest.

ferences

Bellouin, N.; Quaas, J.; Gryspeerdt, E.; Kinne, S.; Stier, P.; Watson-Parris, D.; Boucher, O.; Carslaw, K.S.; Christensen, M.; Daniau, A.L.; et al. Bounding Global Aerosol Radiative Forcing of Climate Change. *Remote Sens.* **2020**, *58*, 637. [CrossRef] [PubMed]

Zhang, K.; Zhang, W.; Wan, H.; Rasch, P.J.; Ghan, S.J.; Easter, R.C.; Shi, X.; Wang, Y.; Wang, H.; Ma, P.L.; et al. Effective Radiative Forcing of Anthropogenic Aerosols in E3SM Version 1: Historical Changes, Causality, Decomposition, and Parameterization Sensitivities. *Atmos. Chem. Phys.* **2022**, *22*, 9129–9160. [CrossRef]

Lohmann, U.; Feichter, J. Global Indirect Aerosol Effects: A Review. *Atmos. Chem. Phys.* **2005**, *5*, 715–737. [CrossRef]

Dubovik, O.; King, M.D. A Flexible Inversion Algorithm for Retrieval of Aerosol Optical Properties from Sun and Sky Radiance Measurements. *J. Geophys. Res. Atmos.* **2000**, *105*, 20673–20696. [CrossRef]

Nakayama, T.; Zhao, W.; Zhang, W. Measurements of Aerosol Optical Properties Using Spectroscopic Techniques. In *Advances in Spectroscopic Monitoring of the Atmosphere*; Elsevier: Amsterdam, The Netherlands, 2021; pp. 345–412. [CrossRef]

Brasil, A.M.; Farias, T.L.; Carvalho, M.G. A Recipe for Image Characterization of Fractal-like Aggregates. *J. Aerosol Sci.* **1999**, *30*, 1113–1125. [CrossRef]

Oh, C.; Sorensen, C.M. The Effect of Overlap between Monomers on the Determination of Fractal Cluster Morphology. *J. Colloid Interface Sci.* **1997**, *193*, 121–130. [CrossRef] [PubMed]

Xiong, C.; Friedlander, S.K. Morphological Properties of Atmospheric Aerosol Aggregates. *Proc. Natl. Acad. Sci. USA* **2001**, *98*, 5825–5829. [CrossRef]

Pang, Y.; Wang, Y.; Wang, Z.; Zhang, Y.; Liu, L.; Kong, S.; Liu, F.; Shi, Z.; Li, W. Quantifying the Fractal Dimension and Morphology of Individual Atmospheric Soot Aggregates. *J. Geophys. Res. Atmos.* **2022**, *127*. [CrossRef]

Dye, A.L.; Rhead, M.M.; Trier, C.J. The Quantitative Morphology of Roadside and Background Urban Aerosol in Plymouth, UK. *Atmos. Environ.* **2000**, *34*, 1617–1624. [CrossRef]

DeCarlo, P.F.; Slowik, J.G.; Worsnop, D.R.; Davidovits, P.; Jimenez, J.L. Particle Morphology and Density Characterization by Combined Mobility and Aerodynamic Diameter Measurements. Part 1: Theory. *Aerosol Sci. Technol.* **2004**, *38*, 1185–1205. [CrossRef]

Carabali, G.; Castro, T.; De La Cruz, W.; Peralta, O.; Varela, A.; Amelines, O.; Rivera, M.; Ruiz-Suarez, G.; Torres-Jardón, R.; Martines-Quiroz, E.; et al. Morphological and Chemical Characterization of Soot Emitted during Flaming Combustion Stage of Native-Wood Species Used for Cooking Process in Western Mexico. *J. Aerosol Sci.* **2016**, *95*, 1–14. [CrossRef]

Carabali, G.; Mamani-Paco, R.; Castro, T.; Peralta, O.; Herrera, E.; Trujillo, B. Optical Properties, Morphology and Elemental Composition of Atmospheric Particles at T1 Supersite on MILAGRO Campaign. *Atmos. Chem. Phys.* **2012**, *12*, 2747–2755. [CrossRef]

Zhang, Y.; Li, Z.; Zhang, Y.; Li, D.; Qie, L.; Che, H.; Xu, H. Estimation of Aerosol Complex Refractive Indices for Both Fine and Coarse Modes Simultaneously Based on AERONET Remote Sensing Products. *Atmos. Meas. Tech.* **2017**, *10*, 3203–3213. [CrossRef]

Smirnov, A.; Holben, B.N.; Eck, T.F.; Dubovik, O.; Slutsker, I. Cloud-Screening and Quality Control Algorithms for the AERONET Database. *Remote Sens. Environ.* **2000**, *73*, 337–349. [CrossRef]

Dubovik, O.; Holben, B.; Eck, T.F.; Smirnov, A.; Kaufman, Y.J.; King, M.D.; Tanré, D.; Slutsker, I. Variability of Absorption and Optical Properties of Key Aerosol Types Observed in Worldwide Locations. *J. Atmos. Sci.* **2002**, *59*, 590–608. [CrossRef]

Moffet, R.C.; Prather, K.A. In-Situ Measurements of the Mixing State and Optical Properties of Soot with Implications for Radiative Forcing Estimates. *Proc. Natl. Acad. Sci. USA* **2009**, *106*, 11872–11877. [CrossRef]

Holben, B.N.; Eck, T.F.; Slutsker, I.; Tanré, D.; Buis, J.P.; Setzer, A.; Vermote, E.; Reagan, J.A.; Kaufman, Y.J.; Nakajima, T.; et al. AERONET—A Federated Instrument Network and Data Archive for Aerosol Characterization. *Remote Sens. Environ.* **1998**, *66*, 1–16. [CrossRef]

Holben, B.N.; Tanré, D.; Smirnov, A.; Eck, T.F.; Slutsker, I.; Abuhassan, N.; Newcomb, W.W.; Schafer, J.S.; Chatenet, B.; Lavenu, F.; et al. An Emerging Ground-Based Aerosol Climatology: Aerosol Optical Depth from AERONET. *J. Geophys. Res. Atmos.* **2001**, *106*, 12067–12097. [CrossRef]

Carabali, G.; Estévez, H.R.; Valdés-Barrón, M.; Bonifaz-Alfonzo, R.; Riveros-Rosas, D.; Velasco-Herrera, V.M.; Vázquez-Gálvez, F.A. Aerosol Climatology over the Mexico City Basin: Characterization of Optical Properties. *Atmos. Res.* **2017**, *194*, 190–201. [CrossRef]

Mora, M.; Braun, R.A.; Shingler, T.; Sorooshian, A. Analysis of Remotely Sensed and Surface Data of Aerosols and Meteorology for the Mexico Megalopolis Area between 2003 and 2015. *J. Geophys. Res. Atmos.* **2017**, *122*, 8705–8723. [CrossRef]

De Foy, B.; Varela, J.R.; Molina, L.T.; Molina, M.J. Rapid Ventilation of the Mexico City Basin and Regional Fate of the Urban Plume. *Atmos. Chem. Phys.* **2006**, *6*, 2321–2335. [CrossRef]

Jáuregui, E. The Climate of the Mexico City Air Basin: Its Effects on the Formation and Transport of Pollutants. In *Urban Air Pollution and Forests*; Fenn, M.E., de Bauer, L.I., Hernández-Tejeda, T., Eds.; Ecological Studies; Springer: New York, NY, USA, 2002; Volume 156, pp. 86–117. [CrossRef]

Marple, V.A.; Rubow, K.L.; Behm, S.M. A Microorifice Uniform Deposit Impactor (Moudi): Description, Calibration, and Use. *Aerosol Sci. Technol.* **1991**, *14*, 434–436. [CrossRef]

25. Carabali, G.; Villanueva-Macias, J.; Ladino, L.A.; Álvarez-Ospina, H.; Raga, G.B.; Andraca-Ayala, G.; Miranda, J.; Grutter, Silva, M.M.; Riveros-Rosas, D. Characterization of Aerosol Particles during a High Pollution Episode over Mexico City. *Sci.* **2021**, *11*, 22533. [CrossRef] [PubMed]
26. Mandelbrot, B. How Long Is the Coast of Britain? Statistical Self-Similarity and Fractional Dimension. *Science* **1967**, *156*, 636– [CrossRef] [PubMed]
27. Mamani-Paco, R.M.; Helble, J.J. Particle Size and Time of the Day Influences on the Morphology Distributions of Atmosphe Fine Particles at the Baltimore Supersite. *Atmos. Environ.* **2007**, *41*, 8021–8029. [CrossRef]
28. Eck, T.F.; Holben, B.N.; Reid, J.S.; Mukelabai, M.M.; Piketh, S.J.; Torres, O.; Jethva, H.T.; Hyer, E.J.; Ward, D.E.; Dubovik, O.; e A Seasonal Trend of Single Scattering Albedo in Southern African Biomass-Burning Particles: Implications for Satellite Produ and Estimates of Emissions for the World's Largest Biomass-Burning Source. *J. Geophys. Res. Atmos.* **2013**, *118*, 6414–64 [CrossRef]
29. Gobbi, G.P.; Kaufman, Y.J.; Koren, I.; Eck, T.F. Classification of Aerosol Properties Derived from AERONET Direct Sun D *Atmos. Chem. Phys.* **2007**, *7*, 453–458. [CrossRef]
30. Rios, B.; Díaz-Esteban, Y.; Raga, G.B. Smoke Emissions from Biomass Burning in Central Mexico and Their Impact on Air Qua in Mexico City: May 2019 Case Study. *Sci. Total Environ.* **2023**, *904*, 166912. [CrossRef]
31. Retama, A.; Ramos-Cerón, M.; Rivera-Hernández, O.; Allen, G.; Velasco, E. Aerosol Optical Properties and Brown Carbon Mexico City. *Environ. Sci. Atmos.* **2022**, *2*, 315–334. [CrossRef]
32. Chew, B.N.; Campbell, J.R.; Reid, J.S.; Giles, D.M.; Welton, E.J.; Salinas, S.V.; Liew, S.C. Tropical Cirrus Cloud Contamination Sun Photometer Data. *Atmos. Environ.* **2011**, *45*, 4041–4058. [CrossRef]
33. Raga, G.B.; Baumgardner, D.; Castro, T.; Martínez-Arroyo, A.; Navarro-González, R. Mexico City Air Quality: A Qualitat Review of Gas and Aerosol Measurements (1960–2000). *Atmos. Environ.* **2001**, *35*, 553–562. [CrossRef]
34. Kutralam-Muniasamy, G.; Pérez-Guevara, F.; Roy, P.D.; Elizalde-Martínez, I.; Shruti, V.C. Impacts of the COVID-19 Lockdo on Air Quality and Its Association with Human Mortality Trends in Megapolis Mexico City. *Air Qual. Atmos. Health* **2021**, 553–562. [CrossRef] [PubMed]
35. Saha, A.; Moorthy, K.K. Impact of Precipitation on Aerosol Spectral Optical Depth and Retrieved Size Distributions: A C Study. *J. Appl. Meteorol.* **2004**, *43*, 902–914. [CrossRef]
36. Moffet, R.C.; de Foy, B.; Molina, L.T.; Molina, M.J.; Prather, K.A. Measurement of ambient aerosols in northern Mexico City single particle mass spectrometry. *Atmos. Chem. Phys.* **2008**, *8*, 4499–4516. [CrossRef]
37. Karydis, V.A.; Tsimpidi, A.P.; Lei, W.; Molina, L.T.; Pandis, S.N. Formation of Semivolatile Inorganic Aerosols in the Mexico C Metropolitan Area during the MILAGRO Campaign. *Atmos. Chem. Phys.* **2011**, *11*, 13305–13323. [CrossRef]
38. Wagner, R.; Ajtai, T.; Kandler, K.; Lieke, K.; Linke, C.; Müller, T.; Schnaiter, M.; Vragel, M. Complex Refractive Indices of Saha Dust Samples at Visible and near UV Wavelengths: A Laboratory Study. *Atmos. Chem. Phys.* **2012**, *12*, 2491–2512. [CrossRef]
39. Wang, S.; Fang, L.; Gu, X.; Yu, T.; Gao, J. Comparison of Aerosol Optical Properties from Beijing and Kanpur. *Atmos. Envir* **2011**, *45*, 7406–7414. [CrossRef]
40. Li, Z.; Gu, X.; Wang, L.; Li, D.; Xie, Y.; Li, K.; Dubovik, O.; Schuster, G.; Goloub, P.; Zhang, Y.; et al. Aerosol Physical and Chemi Properties Retrieved from Ground-Based Remote Sensing Measurements during Heavy Haze Days in Beijing Winter. *Atn Chem. Phys.* **2013**, *13*, 10171–10183. [CrossRef]
41. Cheng, T.; Xu, C.; Duan, J.; Wang, Y.; Leng, C.; Tao, J.; Che, H.; He, Q.; Wu, Y.; Zhang, R.; et al. Seasonal Variation and Difference Aerosol Optical Properties in Columnar and Surface Atmospheres over Shanghai. *Atmos. Environ.* **2015**, *123*, 315–326. [CrossR
42. McConnell, C.L.; Formenti, P.; Highwood, E.J.; Harrison, M.A.J. Using Aircraft Measurements to Determine the Refractive Inc of Saharan Dust during the DODO Experiments. *Atmos. Chem. Phys.* **2010**, *10*, 3081–3098. [CrossRef]
43. Greed, G.; Haywood, J.M.; Milton, S.; Keil, A.; Christopher, S.A.; Gupta, P.; Highwood, E.J. Aerosol Optical Depths over No Africa: 2. Modeling and Model Validation. *J. Geophys. Res. Atmos.* **2008**, *113*, D00C005. [CrossRef]
44. Raut, J.C.; Chazette, P. Vertical Profiles of Urban Aerosol Complex Refractive Index in the Frame of ESQUIF Airborne Measu ments. *Atmos. Chem. Phys.* **2008**, *8*, 901–919. [CrossRef]
45. Beekmann, M.; Prévôt, A.S.H.; Drewnick, F.; Sciare, J.; Pandis, S.N.; Denier Van Der Gon, H.A.C.; Crippa, M.; Freutel, F.; Poula L.; Ghersi, V.; et al. In Situ, Satellite Measurement and Model Evidence on the Dominant Regional Contribution to Fine Particul Matter Levels in the Paris Megacity. *Atmos. Chem. Phys.* **2015**, *15*, 9577–9591. [CrossRef]
46. Tariq, S.; ul-Haq, Z.; Nawaz, H.; Mehmood, U.; Babar, Z. Remote Sensing of Aerosols Due to Biomass Burning over Kanp Sao-Paulo, Ilorin, and Canberra. *J. Atmos. Chem.* **2023**, *80*, 1–52. [CrossRef]

atmosphere

MDPI

icle

Surprisingly High Enhancing Potential of Nitric Acid in ulfuric Acid–Methylamine Nucleation

kang Qiao, Rongjie Zhang, Qiaojing Zhao, Fangfang Ma, Jingwen Chen and Hong-Bin Xie *

Key Laboratory of Industrial Ecology and Environmental Engineering (Ministry of Education), School of Environmental Science and Technology, Dalian University of Technology, Dalian 116024, China; 151676679@mail.dlut.edu.cn (F.Q.); zhangrongjie@mail.dlut.edu.cn (R.Z.); zhaoqiaojing@mail.dlut.edu.cn (Q.Z.); maff@dlut.edu.cn (F.M.); jwchen@dlut.edu.cn (J.C.)

* Correspondence: hbxie@dlut.edu.cn; Tel./Fax: +86-411-84707251

Abstract: Nitric acid (NA) has recently been found to enhance sulfuric acid (SA)-driven new particle formation (NPF) at low temperatures (≤240 K). However, studies on the role of NA in atmospheric NPF remain limited. Herein, we explored the enhancement effect of NA on binary SA–methylamine (MA) nucleation by investigating the mechanism and kinetics of $(NA)_x(SA)_y(MA)_z$ ($0 \leq x$, $0 \leq y$, $x + y \leq 3$, $0 \leq z \leq 3$) clusters using quantum chemical calculations and cluster dynamics simulations. We found that the mixed ternary NA-SA-MA clusters have lower evaporation rates compared to the corresponding NA-SA–dimethylamine (DMA) and NA-SA–ammonia (A) clusters, indicating the stronger binding ability of NA with respect to SA-MA clusters. At atmospheric conditions (T ≥ 278.15 K), NA can enhance the cluster formation rate of SA-MA by about six orders of magnitude, demonstrating a surprisingly high enhancing potential. Moreover, NA acts as an important participant in the cluster growth pathways of the NA-SA-MA system, as opposed to the "bridging" role of NA in the previously studied NA-SA-A system. This study proposes the first case of NA efficiently enhancing SA–amine nucleation at ambient temperature, suggesting a larger impact of NA in atmospheric NPF than previously expected.

Keywords: particle formation; nitric acid; sulfuric acid; methylamine; quantum chemical calculation; atmospheric cluster dynamics code

1. Introduction

Atmospheric particulate matter poses a serious threat to global climate and air quality [1–6]. New particle formation (NPF) plays a pivotal role as an important contributor to atmospheric particulate matter, which can contribute over 50% of the global aerosol budget [1–12]. Therefore, a comprehensive understanding of NPF is essential for evaluating the impacts of particulate matter and devising effective control strategies.

The most important nucleation precursor of atmospheric NPF is thought to be sulfuric acid owing to its exceptionally low volatility and strong acidity [3,6,13–24]. Atmospheric bases, such as ammonia (A, NH_3) and amines, can efficiently nucleate with SA via acid–base proton transfer reactions and the formation of hydrogen bonds [3,6,13,14,18,19,25–31]. Despite large differences in concentrations, amines have been found to more efficiently enhance SA particle formation compared with A [13,14,24,28,32]. However, the binary SA–base nucleation mechanisms fall short of explaining the high rates of NPF observed under realistic atmospheric conditions [33–35]. This implies that other potential gaseous precursors may participate in and further enhance binary SA–base nucleation

Nitric acid (NA, HNO_3) is an abundant inorganic acid in the troposphere, with a broad variety of sources and a wide range of distributions [36–42]. The tropospheric concentration of NA can be up to 1×10^{12} molecules cm^{-3}, significantly exceeding that of SA by a factor of four to seven orders of magnitude [40,43–45]. Considering the high atmospheric abundance of NA, some recent investigations have concentrated on the enhancing effect of NA on

SA–base nucleation [46–51]. Wang et al. and Liu et al. pointed out that NA can effectively enhance both SA-DMA and SA-A nucleation at low temperatures ($\leq$240 K), but NA had a negligible enhancing effect at temperatures commonly encountered in the atmospheric boundary layer (280 K) [46–48]. In addition, Knattrup et al. found that NA can increase the cluster formation rate by studying mixed-acid–mixed-base clusters containing at most one NA molecule [49,50]. Amines cover a wide variety of species in the atmosphere with different basicities and structures, implying that the enhancing effect of NA may be different for different SA–amine systems. However, except for DMA and A, the enhancement effect of NA on SA–other amine nucleation has not been studied in detail, which hinders a comprehensive assessment of the role of NA in NPF processes. Therefore, it is essential to explore the enhancing effect of NA on SA–other amine nucleation in detail.

Methylamine (MA) has a higher global emission than DMA and a stronger basicity than A [52]. In addition, MA has the structural characteristics of a lower steric hindrance than DMA and the capability of forming one more hydrogen bond. Previous studies have shown that basicity, hydrogen-bonding capacity and steric hindrance jointly influence the clustering ability of a given amine with acids [53–56]. This implies that the enhancing effect of NA on SA-MA nucleation is not easy to predict simply based on the enhancing effect of NA on the previously studied SA-DMA and SA-A systems. Here, we conducted a systematic investigation of NA's enhancing effect on SA-MA nucleation by exploring the thermodynamics and kinetics of the ternary NA-SA-MA system using quantum chemical calculations with simulations using the Atmospheric Cluster Dynamic Code (ACDC) [57]. Additionally, we chose the previously studied NA-SA-DMA and NA-SA-A systems as references to analyze and compare the enhancing role of NA in these three systems [47,48].

2. Materials and Methods

2.1. Configurational Sampling

Similar to our previous studies [53,58–60], a multi-step sampling approach was utilized to explore the most stable configurations of the clusters in the ternary NA-SA-MA system, including $(NA)_{1\text{-}3}(MA)_{1\text{-}3}$ and $(NA)_x(SA)_y(MA)_z$ ($1 \leq x$, $1 \leq y$, $x + y \leq 3$, $1 \leq z \leq 3$) clusters. The configurations of the remaining clusters in the ternary NA-SA-MA system ($(SA)_{0\text{-}3}(MA)_{0\text{-}3}$ and homomolecular $(NA)_{1\text{-}3}$) were derived from previous research [48,61]. In order to obtain the most stable configurations, the ABCluster program was first employed to generate 6000 initial guess structures for each cluster type [62,63]. The CHARMM36 force field was applied for the simulation of all molecules in the ABCluster program. Based on the initial guess structures, four sequential calculation steps were performed to gradually screen for the global minimum. The specific details of the multi-step sampling approach and the employed calculation methods are presented in Appendix A. In brief, the configuration exhibiting the lowest Gibbs free energy value (G) was identified as the most stable configuration for a given cluster type. The binding free energy (ΔG) was derived from the subtraction of the individual G of the constituent monomers from the overall G of the studied cluster at a temperature of 298.15 K. All calculations utilizing the PM6 and ωB97X-D methods were performed through the GAUSSIAN 09 software [64], while the DLPNO-CCSD(T)/aug-cc-pVTZ computations were carried out by the ORCA 4.0.0 software [65]. In addition, all the cluster configurations for the ternary NA-SA-DMA and NA-SA-A systems, as reported in earlier research [47,48], were adopted and subsequently recalculated at the theoretical levels utilized in the current investigation.

2.2. Atmospheric Cluster Dynamics Code Simulations

ACDC was utilized to investigate the time evolution of cluster concentrations and growth pathways. For more details about the theory behind ACDC, see Appendix A [57,66]. Here, the simulated ternary NA-SA-MA system can be characterized as a "3×3" box involving $(NA)_x(SA)_y(MA)_z$ ($1 \leq x$, $1 \leq y$, $x + y \leq 3$, $1 \leq z \leq 3$), $(NA)_{0\text{-}3}(MA)_{0\text{-}3}$ and $(SA)_{0\text{-}3}(MA)_{0\text{-}3}$ clusters. The choice of boundary clusters was guided by their stability at the specific temperatures examined, as detailed in Appendix A. Within the simulations, the SA concentration [SA] was fixed in the range of 10^6 to 10^7 cm^{-3}, the NA concentration [NA] was adjusted between 10^8 and 10^{12} cm^{-3}, and the MA concentration [MA] was set between 10^8 and 10^9 cm^{-3}, which values correspond to their usual ambient concentrations [3,24, 39]. The simulations were conducted at 238.15 K to 298.15 K, and the coefficient for the coagulation sink was established at 2×10^{-2} s^{-1}, which corresponds to the typical value for polluted urban atmospheres [3]. As a comparison, the NA-SA-DMA and NA-SA-A systems were also simulated, with the DMA concentration [DMA] between 10^8 and 10^9 cm^{-3} and the A concentration [A] at 10^{10} cm^{-3}. The simulation system for both the NA-SA-DMA and the NA-SA-A systems also consisted of a "3 × 3" box, and the selection of the boundary clusters can be seen in Appendix A. All other conditions were set the same as those used for the simulation of the NA-SA-MA system.

3. Results and Discussion

3.1. Cluster Structures

Given that $(SA)_{1\text{-}3}(MA)_{1\text{-}3}$ clusters have been extensively discussed in earlier studies [24,26,66], here, we primarily concentrate on the binary $(NA)_{1\text{-}3}(MA)_{1\text{-}3}$ and mixed ternary $(NA)_x(SA)_y(MA)_z$ ($1 \leq x, 1 \leq y, x + y \leq 3, 1 \leq z \leq 3$) clusters. The most stable conformations of the $(NA)_x(SA)_y(MA)_z$ ($1 \leq x, 1 \leq y, x + y \leq 3, 1 \leq z \leq 3$) clusters are shown in Figure 1, and the most stable conformations of the $(NA)_{1\text{-}3}(MA)_{1\text{-}3}$ clusters are displayed in Figure S1. Proton transfers from NA to MA occur in all the binary $(NA)_{1\text{-}3}(MA)_{1\text{-}3}$ clusters except $(NA)_1(MA)_1$, and these acid–base clusters are stabilized through electrostatic interactions between positively and negatively charged ions and intermolecular hydrogen bonds. The $(NA)_1(MA)_1$ cluster is the exception, as there is no proton transfer; it is only stabilized by hydrogen bonds. For all the mixed ternary $(NA)_x(SA)_y(MA)_z$ ($1 \leq x, 1 \leq y, x + y \leq 3, 1 \leq z \leq 3$) clusters, proton transfer and the formation of a complex hydrogen bond network are observed. When $y \geq z$, all MA molecules undergo protonation by accepting a proton from SA, such as $(NA)_1(SA)_1(MA)_1$, $(NA)_1(SA)_2(MA)_1$, $(NA)_2(SA)_1(MA)_1$ and $(NA)_1(SA)_2(MA)_2$. When $y < z$, both SA and NA can transfer a proton to the MA molecules. Thus, the proton transfer process from SA to MA is comparatively more advantageous compared to the transfer from NA to MA in the case of all the mixed ternary NA-SA-MA clusters. This phenomenon is likely caused by the higher acidity of SA compared to NA. In addition, only one proton can be transferred to MA from SA molecules in all the acid–base clusters, and therefore no formation of SO_4^{2-} is observed.

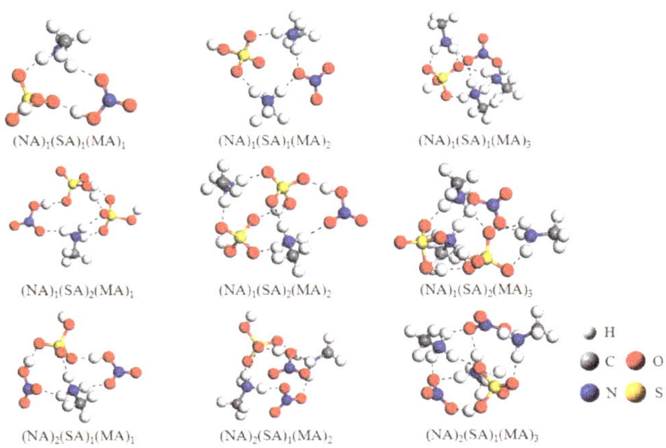

Figure 1. The most stable configurations of the $(NA)_x(SA)_y(MA)_z$ ($1 \leq x, 1 \leq y, x + y \leq 3, 1 \leq z \leq 3$) clusters calculated at the DLPNO-CCSD(T)/aug-cc-pVTZ//ωB97X-D/6-31++G(d,p) level of theory at 298.15 K. The dashed black lines represent hydrogen bonds.

We then compared the number of hydrogen bonds within the mixed ternary $(NA)_x(SA)_y(MA)_z$ cluster with the numbers in the $(NA)_x(SA)_y(DMA)_z$ and $(NA)_x(SA)_y(A)_z$ clusters ($1 \leq x$, $1 \leq y$, $x + y \leq 3$, $1 \leq z \leq 3$) (see Table 1), since the hydrogen bond is a vital structural factor for stabilizing clusters. It was found that the numbers of hydrogen bonds in all the $(NA)_x(SA)_y(MA)_z$ clusters were greater than or equal to those in the corresponding $(NA)_x(SA)_y(DMA)_z$ clusters, except for the case of $(NA)_1(SA)_1(MA)_1$ and $(NA)_1(SA)_1(MA)_2$. This is related to the presence of more hydrogen bond formation sites in MA than DMA, as it has one more methyl group. In addition, the numbers of hydrogen bonds in the $(NA)_x(SA)_y(MA)_z$ clusters are comparable to those in the $(NA)_x(SA)_y(A)_z$ clusters. Hence, the observed numbers of hydrogen bonds in the mixed ternary $(NA)_x(SA)_y(MA)_z$ clusters suggest that they are likely more stable.

Table 1. The numbers of hydrogen bonds (n) in $(NA)_x(SA)_y(MA)_z$, $(NA)_x(SA)_y(DMA)_z$ and $(NA)_x(SA)_y(A)_z$ ($1 \leq x, 1 \leq y, 1 \leq z \leq x + y \leq 3$) clusters.

Clusters	n	Clusters	n	Clusters	n
$(NA)_1(SA)_1(MA)_1$	3	$(NA)_1(SA)_1(DMA)_1$	4	$(NA)_1(SA)_1(A)_1$	3
$(NA)_1(SA)_1(MA)_2$	4	$(NA)_1(SA)_1(DMA)_2$	5	$(NA)_1(SA)_1(A)_2$	5
$(NA)_2(SA)_1(MA)_1$	5	$(NA)_2(SA)_1(DMA)_1$	5	$(NA)_2(SA)_1(A)_1$	5
$(NA)_2(SA)_1(MA)_2$	7	$(NA)_2(SA)_1(DMA)_2$	6	$(NA)_2(SA)_1(A)_2$	8
$(NA)_2(SA)_1(MA)_3$	10	$(NA)_2(SA)_1(DMA)_3$	7	$(NA)_2(SA)_1(A)_3$	9
$(NA)_1(SA)_2(MA)_1$	6	$(NA)_1(SA)_2(DMA)_1$	5	$(NA)_1(SA)_2(A)_1$	6
$(NA)_1(SA)_2(MA)_2$	7	$(NA)_1(SA)_2(DMA)_2$	7	$(NA)_1(SA)_2(A)_2$	8
$(NA)_1(SA)_2(MA)_3$	11	$(NA)_1(SA)_2(DMA)_3$	8	$(NA)_1(SA)_2(A)_3$	10

3.2. Evaporation Rates

The evaporation rate of a cluster represents the rate at which molecules or smaller clusters escape from the original cluster, and therefore it can be used to infer the stability of the cluster. The evaporation rate of a mother cluster is relevant to the collision rate coefficient of its two daughter clusters or monomers and the difference in formation free energy between the mother cluster and its two possible daughter clusters or monomers (detailed in Equation (A3) in Appendix A). Since the difference between the collision rate coefficients for various systems is minor, the evaporation rate of a mother cluster mostly depends on the relative formation free energy of the mother cluster with respect to its possible daughter clusters or monomers. In general, clusters with lower evaporation rates are considered to be more stable [58–60]. Since we are mainly concerned with the enhancement effect of NA on SA-MA nucleation, here, we primarily focus on the evaporation rate based on the ΔG values (Figure S2) of the mixed ternary NA-SA-MA clusters $(NA)_x(SA)_y(MA)_z$ ($1 \leq x$, $1 \leq y$, $x + y \leq 3$, $1 \leq z \leq 3$). Figure 2A shows the evaporation rates for the mixed ternary NA-SA-MA clusters at 278.15 K, while Figure 2B provides the evaporation rates for the binary SA-MA clusters as a comparison. For the ternary NA-SA-MA system, the evaporation rates of the four clusters $(NA)_1(SA)_2(MA)_3$, $(NA)_2(SA)_1(MA)_3$, $(NA)_1(SA)_2(MA)_2$ and $(NA)_1(SA)_1(MA)_2$ are on the order of 10^{-3}–10^{-2} s^{-1}. These are significantly lower than those of the other ternary NA-SA-MA clusters, which have evaporation rates higher than 10^1 s^{-1}. Under atmospheric conditions, these four clusters may be considered to possess stability. Such a wide variation in the evaporation rate for various clusters is understandable, since the relative formation free energies of mother clusters with respect to their daughter clusters or monomers differ greatly. In comparison with the binary SA-MA clusters, we found that ternary NA-SA-MA clusters with equal acid and base molecule numbers, i.e., $(NA)_1(SA)_2(MA)_3$, $(NA)_2(SA)_1(MA)_3$ and $(NA)_1(SA)_1(MA)_2$, have lower evaporation rates than the corresponding binary SA-MA clusters. This suggests that the addition of NA enhances the stability of these three clusters, and there is a strong possibility that they can participate in the main cluster growth pathways of the NA-SA-MA system.

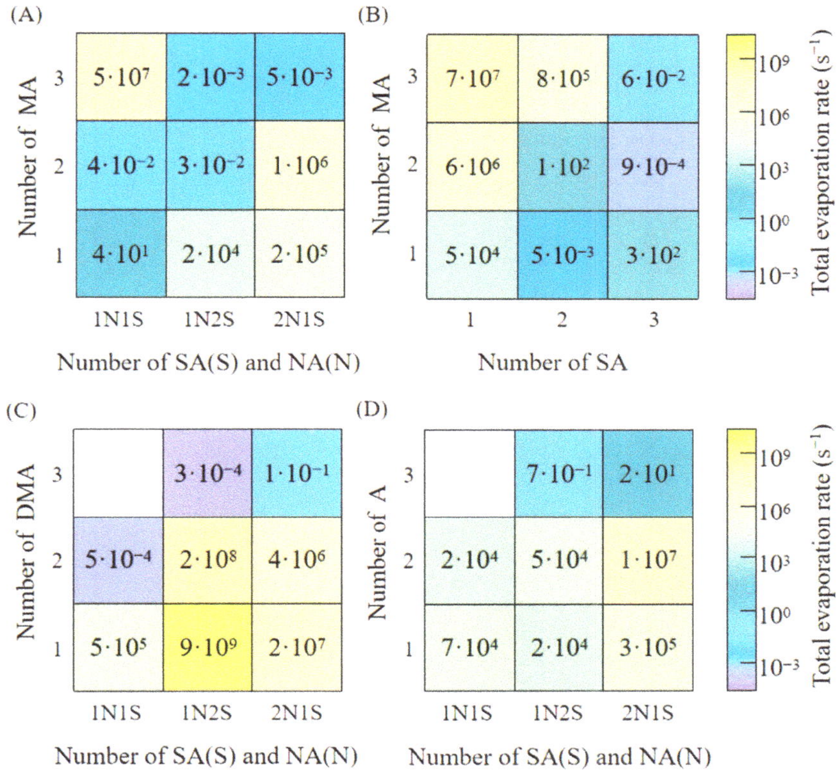

Figure 2. Evaporation rates of the (**A**) $(NA)_x(SA)_y(MA)_z$ ($1 \leq x$, $1 \leq y$, $x + y \leq 3$, $1 \leq z \leq 3$), (**B**) $(SA)_{1\text{-}3}(MA)_{1\text{-}3}$, (**C**) $(NA)_x(SA)_y(DMA)_z$ ($x \geq 1, y \geq 1, 1 \leq z \leq x + y \leq 3$) and (**D**) $(NA)_x(SA)_y(A)_z$ ($x \geq 1, y \geq 1, 1 \leq z \leq x + y \leq 3$) clusters at 278.15 K.

A comparison was also conducted between the evaporation rates of the ternary NA-SA-MA clusters and those of the ternary NA-SA-DMA and NA-SA-A clusters (see Figure 2C,D). A common feature in all three systems is that the evaporation rates for clusters with equivalent numbers of acidic and basic molecules, i.e., $(NA)_1(SA)_1(base)_2$, $(NA)_2(SA)_1(base)_3$ and $(NA)_1(SA)_2(base)_3$, are lower than those for the other ternary clusters. This is consistent with many previous studies [61,67]. Moreover, the evaporation rates of all the ternary NA-SA-MA clusters are lower compared to the corresponding ones in the NA-SA-DMA system, except for the $(NA)_1(SA)_1(MA)_2$ and $(NA)_1(SA)_2(MA)_3$ clusters. It has been observed that the evaporation rate for the $(NA)_1(SA)_2(MA)_2$ cluster is significantly lower compared to that of the $(NA)_1(SA)_2(DMA)_2$ cluster. Similarly, the evaporation rates for each of the ternary NA-SA-MA clusters are lower than those for the related NA-SA-A clusters. Therefore, the lower evaporation rates suggest that ternary NA-SA-MA clusters have greater stability when compared to the ternary NA-SA-DMA and NA-SA-A clusters.

3.3. Addition Free Energies of NA to the Clusters

We employed the addition free energy of NA (ΔG_{add}) as an important parameter to evaluate the binding ability of NA to the binary SA–base clusters. Herein, ΔG_{add} values for the binary SA-MA, SA-DMA and SA-A clusters were calculated to compare the preferences for the uptake of NA between these three binary SA–base systems (Figure 3). The ΔG_{add} for each ternary $(NA)_x(SA)_y(base)_z$ cluster was calculated as follows:

$$\Delta G_{add} = G_{(NA)x(SA)y(base)z} - G_{(SA)y(base)z} - G_{NA} \times x \quad (1)$$

where $G_{(NA)x(SA)y(base)z}$, $G_{(SA)y(base)z}$ and G_{NA} are the Gibbs free energies of the ternary $(NA)_x(SA)_y(base)_z$ cluster, the binary $(SA)_y(base)_z$ cluster and NA monomers, respectively. As shown in Figure 3A, the ΔG_{add} values for all the ternary NA-SA-MA clusters are highly negative, and therefore the addition of NA to the clusters is thermodynamically favorable. Furthermore, when the number of SA molecules stays constant, there is a clear trend that the ΔG_{add} values of the $(NA)_x(SA)_y(MA)_z$ clusters decrease with an increase in the number of base molecules, suggesting that the greater number of bases in the cluster makes it more favorable to attract NA molecules. It is noteworthy that the ΔG_{add} values for all the ternary NA-SA-MA clusters are lower than those for the corresponding ternary NA-SA-DMA and NA-SA-A clusters, signifying that NA binding with SA-MA clusters is more advantageous compared to the SA-DMA and SA-A clusters. We speculate that the introduction of NA leads to a change in the topological structure of the binary SA-MA clusters, resulting in stronger binding of NA with SA-MA clusters and more hydrogen bonds in the formed ternary NA-SA-MA clusters.

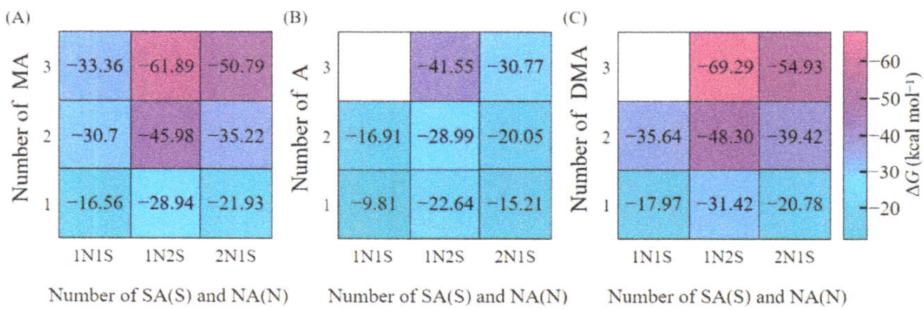

Figure 3. Addition free energy of NA (ΔG_{add}) for the (**A**) $(NA)_x(SA)_y(MA)_z$ ($1 \le x, 1 \le y, x + y \le 3, 1 \le z \le 3$), (**B**) $(NA)_x(SA)_y(DMA)_z$ ($1 \le x, 1 \le y, 1 \le z \le x + y \le 3$) and (**C**) $(NA)_x(SA)_y(A)_z$ ($1 \le x, 1 \le y, 1 \le z \le x + y \le 3$) clusters calculated at the DLPNO-CCSD(T)/aug-cc-pVTZ//ωB97X-D/6-31++G(d,p) level of theory at 298.15 K and 1 atm.

3.4. Enhancement Effect of NA on Cluster Formation Rates

Here, we further investigated the enhancing effect of NA on SA-MA nucleation by comparing the cluster formation rates before and after adding NA to the binary SA-MA system. We also compared it with the enhancing effect of NA on SA-DMA nucleation and SA-A nucleation (cluster formation rates of these nucleation systems are shown in Tables S1–S3). The enhancing effect of NA on binary SA–base nucleation is quantified by the enhancement factor (r):

$$r = \frac{J([SA] = x, [base] = y, [NA] = z)}{J([SA] = x, [base] = y, [NA] = 0)} \quad (2)$$

where J ([SA] = x, [base] = y, [NA] = z) and J ([SA] = x, [base] = y, [NA] = 0) indicate the cluster formation rate of the ternary NA-SA–base system and the binary SA–base system without NA, respectively. Under realistic atmospheric conditions, precursor concentration and temperature are both important factors influencing the nucleation rate. Therefore, we examined r at different precursor concentrations and temperatures.

The effect of temperature on r when [NA] = 10^{10} cm^{-3} and T = 238.15 K–298.15 K is displayed in Figure 4. It is observable that NA always has a significant enhancing effect on SA-MA nucleation, with the r value ranging from 276 to 14370 under the considered conditions. The overall trend shows that the r value increases with increasing temperature, except for the case with the conditions of [SA] = 10^6 cm^{-3} and [MA] = 10^9 cm^{-3}, where the *r* value initially decreases and then increases with increasing temperature. Furthermore, the *r* value under the lower SA concentration (10^6 cm^{-3}) is higher than that under the higher SA concentration (10^7 cm^{-3}), indicating that NA can cause stronger enhancement of SA-MA nucleation when the SA is scarce. Furthermore, for the NA-SA-DMA and NA-SA-A systems, the *r* value increases with decreasing temperature, and NA only has a prominent enhancing effect on binary SA-DMA/A nucleation at temperatures below 258.15 K. However, the *r* value of NA in the binary SA-DMA/A nucleation is much lower than that in the binary SA-MA nucleation under our considered conditions, which confirms the significantly high enhancement effect of NA on SA-MA nucleation.

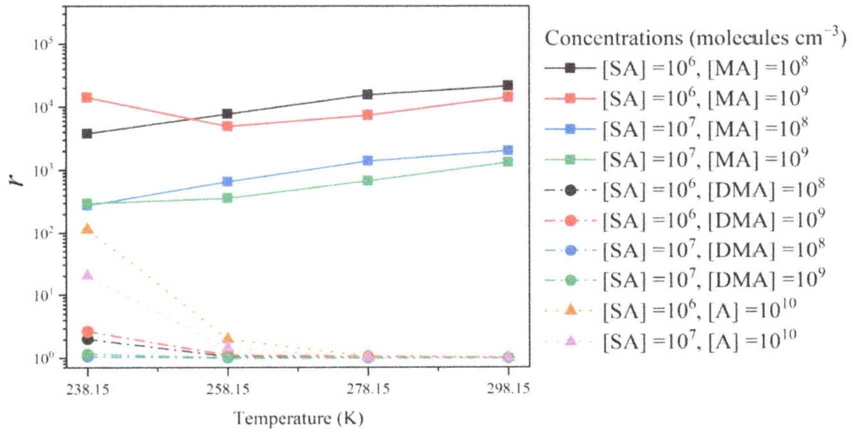

Figure 4. Enhancement factor (*r*) of NA in SA-MA, SA-DMA and SA-A nucleation at varying temperatures and precursor concentrations. The concentration of NA is fixed at 10^{10} cm^{-3}. The solid black, red, blue and green lines represent the NA-SA-MA system; the dashed–dotted black, red, blue and green lines represent the NA-SA-DMA system; and the dotted orange and purple lines represent the NA-SA-A system.

Figure 5 shows the effect of NA concentration on *r* at 278.15 K. It can be seen that for three NA-SA-base systems, *r* shows a positive dependence on [NA] and the *r* value under the lower SA concentration (10^6 cm^{-3}) is higher than that under the higher SA concentration (10^7 cm^{-3}). For the NA-SA-MA system, *r* increases from 16 to about 10^6 as the NA concentration increases from 10^8 cm^{-3} to 10^{12} cm^{-3}, reflecting the high enhancing effect of NA for the SA-MA system at 278.15 K. However, for the NA-SA-DMA and NA-SA-A systems, an enhancing effect of NA was only observed when NA > 10^{10} cm^{-3} at 278.15 K, which is still much lower than the *r* value for the NA-SA-MA system at the same NA concentration.

As discussed above, our results indicate that even under environmental conditions with high temperatures, such as at noon, NA can still efficiently enhance SA-MA nucleation to a surprisingly high degree, in contrast to the negligible enhancing effect of NA on binary SA-DMA or SA-A nucleation above 278.15 K presented in previous studies [46–48].

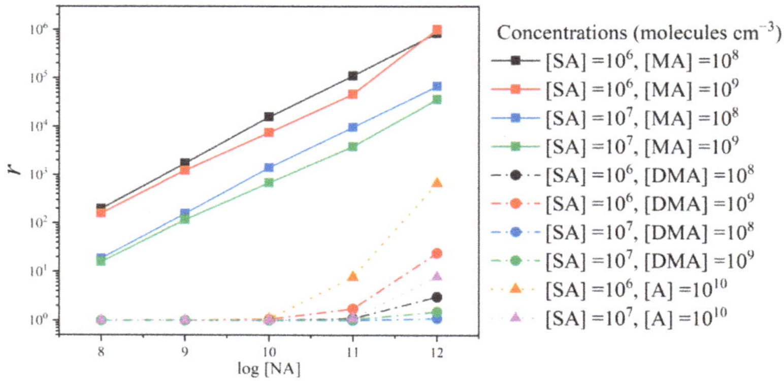

Figure 5. Enhancement factor (r) of NA in SA-MA, SA-DMA and SA-A nucleation at 278.15 K and with varying concentrations of NA (from 10^8 to 10^{12} cm^{-3}). The solid black, red, blue and green lines represent the NA-SA-MA system; the dashed–dotted black, red, blue and green lines represent the NA-SA-DMA system; and the dotted orange and purple lines represent the NA-SA-A system.

3.5. Mechanism of the NA-SA-MA Nucleation System

ACDC simulations can keep track of the fluxes between clusters, which makes analysis of the cluster formation mechanism easy. Herein, the cluster growth pathway within the NA-SA-MA system at 278.15 K, [NA] = 10^{10} cm^{-3}, [SA] = 10^6 cm^{-3} and [MA] = 10^8 cm^{-3} is displayed in Figure 6. Under these conditions, NA led to an enhancement factor of about 15,000 for SA-MA nucleation. As shown in Figure 6, the cluster growth mainly proceeds through two channels. One is the binary SA-MA channel, which is initiated by the collision of two SA molecules or the collision between SA and MA molecules. The other is the NA-participated channel, which is initiated by the collision of NA and MA molecules. In addition, the two channels are linked by the addition of one NA molecule to the $(SA)_1(MA)_1$ and $(SA)_2(MA)_2$ clusters to form the $(NA)_1(SA)_1(MA)_1$ and $(NA)_1(SA)_2(MA)_2$ clusters, respectively. Both channels subsequently grow through collisions with monomers, ultimately producing the $(SA)_3(MA)_3$ and $(NA)_1(SA)_2(MA)_3$ clusters, respectively. It is noted that the ternary NA-SA-MA clusters that participate in the growth pathways contain at most one NA molecule. Finally, $(SA)_3(MA)_3$ and $(NA)_1(SA)_2(MA)_3$ clusters can directly collide with NA and SA molecules, respectively, to form the $(NA)_1(SA)_3(MA)_3$ cluster, which grows out of the "3 × 3 box", with the two paths contributing 15% and 85% of the cluster formation rates, respectively.

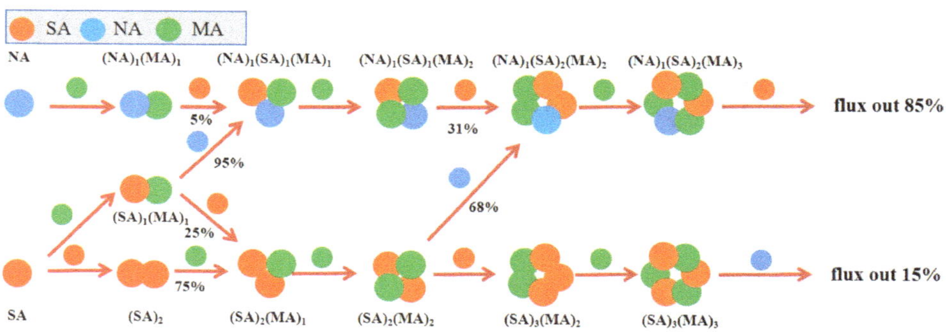

Figure 6. Cluster growth pathways for the NA-SA-MA system at 278.15 K, [NA] = 10^{10} cm^{-3}, [SA] = 10^6 cm^{-3} and [MA] = 10^8 cm^{-3}.

We also tested the influence of temperature and [NA] on the cluster growth pathway of the NA-SA-MA system. Figure S3 presents the cluster growth pathway at a lower temperature (238.15 K, Figure S3A) and a higher [NA] (10^{12} cm^{-3}, Figure S3B). As shown in Figure S3, the main growth pathway is slightly more complicated than that at 278.15 K and [NA] = 10^{10} cm^{-3}. The cluster growth pathway is still mainly divided into the SA-MA channel and the NA-participated channel. However, at 238.15 K or [NA] = 10^{12} cm^{-3}, the two channels can also be connected by an extra route via the evaporation of a NA molecule: $(NA)_1(SA)_1(MA)_1 \rightarrow (NA)_1(SA)_2(MA)_1 \rightarrow (SA)_2(MA)_1$. Another difference is that the finally formed $(NA)_1(SA)_2(MA)_3$ cluster can collide not only with an SA monomer but also with the $(NA)_1(MA)_1$ dimer cluster to grow out of the "3 × 3 box" at 238.15 K or [NA] = 10^{12} cm^{-3}. Additionally, at 238.15 K, $(NA)_1(SA)_1(MA)_2$ can also collide with itself to form the $(NA)_2(SA)_2(MA)_4$ cluster growing out of the "3 × 3 box".

Overall, the growth pathways under the three conditions are generally similar; the NA-participated clusters contribute almost 100% to the cluster formation rate, indicating the importance of the participation and enhancement of NA in SA-MA nucleation. This is different from the case in which NA mainly acts as a "bridge" joining the smaller and larger binary SA-A clusters in the cluster growth pathways of the NA-SA-A system [47].

4. Implications and Conclusions

This study is the first to point out the surprisingly high enhancing effect of NA on the nucleation of the binary SA-MA system at ambient temperature. For example, at [NA] = 10^{10} cm^{-3} and T = 278.15 K–298.15 K, a common concentration and normal temperature range in the troposphere, NA can efficiently enhance the cluster formation rate of the binary SA-MA system by about 10^4-fold. Moreover, the enhancing role of NA could be even more pronounced in local regions with higher concentrations of NA and near emission sources of MA. However, previous studies reported that only at lower temperatures ($\leq$240 K) could NA efficiently enhance binary SA-DMA or SA-A nucleation [47,48]. Thus, the enhancing effect of NA on the investigated binary SA–base nucleation systems does not simply adhere to the sequence of gas basicity among the bases (DMA > MA > A) [68], although acid–base reactions occur between NA and bases in ternary NA-SA–base nucleation. This implies that the enhancing effect of NA on binary SA–base nucleation is related not only to the basicity of the atmospheric bases but also to their structural factors, e.g., their hydrogen-bonding capacity. Therefore, the contribution of NA to nucleation involving SA and other bases warrants further study to comprehensively expand the current understanding of the role of NA in atmospheric NPF.

The enhancing potential of MA in SA-driven nucleation is lower than that of the other typically observed alkyl amines, i.e., DMA and trimethylamine (TMA) [24]. However, recent studies showed that MA can synergistically enhance SA-driven nucleation with other amines [69,70]. Considering the high enhancing effect of NA on SA-MA nucleation, it is likely that NA could significantly enhance the nucleation of SA–multiple amines containing MA in realistic atmospheres. It is important to mention that a recent investigation has revealed that NA can increase the nucleation rate of complex mixed-acid–mixed-base systems (e.g., the quaternary NA-SA-MA-DMA system), but the multi-component clusters in this study only include one NA molecule [50]. Such an NA-enhanced multi-component nucleation mechanism containing more NA and base molecules may explain the intense NPF events in the urban atmosphere, which deserve more concern in the future.

Supplementary Materials: The following supporting information can be downloaded at: https://www.mdpi.com/article/10.3390/atmos15040467/s1, Figure S1: Lowest Gibbs free energy conformations of the $(NA)_{1\text{-}3}(MA)_{1\text{-}3}$ clusters; Figure S2: Formation free energy (ΔG) of (A) $(NA)_x(SA)_y(MA)_z$ ($1 \leq x, 1 \leq y, x + y \leq 3, 1 \leq z \leq 3$), (B) $(NA)_x(SA)_y(A)_z$ ($x \geq 1, y \geq 1, 1 \leq z \leq x + y \leq 3$) and (C) $(NA)_x(SA)_y(DMA)_z$ ($x \geq 1, y \geq 1, 1 \leq z \leq x + y \leq 3$) clusters at 298.15 K; Figure S3: Cluster growth pathways for the NA-SA-MA system at (A) 238.15 K, [NA] = 10^{10} molecules cm^{-3}, [SA] = 10^6 molecules cm^{-3} and [MA] = 10^8 molecules cm^{-3} and (B) 278.15 K, [NA] = 10^{12} molecules cm^{-3}, [SA] = 10^6 molecules cm^{-3} and [MA] = 10^8 molecules cm^{-3}; Table S1: Cluster formation rates (J, cm^{-3} s^{-1}) of the NA-SA-MA system and the SA-MA system, and the enhancement factor (r) at different temperatures (K) and precursor concentrations (molecules cm^{-3}); Table S2: Cluster formation rates (J, cm^{-3} s^{-1}) of the NA-SA-DMA system and the SA-DMA system, and the enhancement factor (r) at different temperatures (K) and precursor concentrations (molecules cm^{-3}); Table S3: Cluster formation rates (J, cm^{-3} s^{-1}) of the NA-SA-A system and the SA-A system, and the enhancement factor (r) at different temperatures (K) and precursor concentrations (molecule cm^{-3}); Table S4: Coordinates of all optimized $(NA)_x(SA)_y(MA)_z$ ($1 \leq x, 1 \leq y, x + y \leq 3, 1 \leq z \leq 3$), $(NA)_x(SA)_y(DMA)_z$ ($1 \leq x, 1 \leq y, 1 \leq z \leq x + y \leq 3$) and $(NA)_x(SA)_y(A)_z$ ($1 \leq x, 1 \leq y, 1 \leq z \leq x + y \leq 3$) clusters; Table S5: Single-point energies (kcal mol^{-1}) at the DLPNO-CCSD(T)/aug-cc-pVTZ level and the corresponding Gibbs free energy correction G_{corr} (kcal mol^{-1}) and Gibbs free energy G (kcal mol^{-1}) at 298.15 K for all the most stable configurations of the NA-SA-MA, NA-SA-DMA and NA-SA-A systems.

Author Contributions: Conceptualization, F.Q. and H.-B.X.; methodology, F.Q. and Q.Z.; formal analysis, F.Q. and R.Z.; investigation, F.Q.; writing—original draft preparation, F.Q.; writing—review and editing, F.Q., Q.Z., R.Z., H.-B.X., F.M. and J.C. All authors have read and agreed to the published version of the manuscript.

Funding: This research was funded by the National Key Research and Development Program of China (2022YFC3701000, Task1) and the National Natural Science Foundation of China (22236004 and 22176022).

Institutional Review Board Statement: Not applicable.

Informed Consent Statement: Not applicable.

Data Availability Statement: The data presented in this study are available in Supplementary Materials of this article.

Conflicts of Interest: The authors declare no conflicts of interest.

Appendix A

Appendix A.1. Specific Details of Multi-Step Sampling Approach and Employed Calculation Methods

The multi-step sampling approach was employed to search for the most stable configurations of the clusters in the ternary NA-SA-MA system, including $(NA)_{1\text{-}3}(MA)_{1\text{-}3}$ and $(NA)_x(SA)_y(MA)_z$ ($1 \leq x, 1 \leq y, x + y \leq 3, 1 \leq z \leq 3$) clusters. First, the ABCluster program was employed to generate 6000 initial guess structures for each cluster type. The CHARMM36 force field was applied for the simulation of all molecules in the ABCluster program. Based on the initial guess structure, four sequential calculation steps were performed, including preliminary geometry optimization at the PM6 level of theory, followed by a single-point energy calculation at the ωB97X-D/6-31+G(d,p) level. Subsequently, a more precise geometry optimization and vibrational frequency analysis were conducted at the ωB97X-D/6-31++G(d,p) level of theory. The final step involved a single-point energy calculation at the DLPNO-CCSD(T)/aug-cc-pVTZ level of theory using the keywords "tightscf", "tightpno", "GRID4" and "nofinalgrid". The Gibbs free energy (G) for each cluster was calculated by combining the single-point energy at the DLPNO-CCSD(T)/aug-cc-pVTZ theoretical level with a correction term for the Gibbs free energy derived from the ωB97X-D/6-31++G(d,p) level of theory. The configuration exhibiting the lowest G value was identified as the most stable configuration for a given cluster type. The binding free energy (ΔG) was derived from the subtraction of the individual G of the constituent monomers from the overall G of the studied cluster at a temperature of 298.15 K. ΔG values at different temperatures were obtained on the premise that both the enthalpy (ΔH) and entropy (ΔS) changes were held constant across the range of temperatures considered, which spanned from 238.15 K to 298.15 K. The coordinates of all the optimized $(NA)_x(SA)_y(MA)_z$ ($1 \leq x, 1 \leq y, x + y \leq 3, 1 \leq z \leq 3$), $(NA)_x(SA)_y(DMA)_z$ ($1 \leq x, 1 \leq y, 1 \leq z \leq x + y \leq 3$) and $(NA)_x(SA)_y(A)_z$ ($1 \leq x, 1 \leq y, 1 \leq z \leq x + y \leq 3$) clusters can be seen in Table S4. Additionally, the single-point energies at the DLPNO-CCSD(T)/aug-cc-pVTZ level and the corresponding Gibbs free energy correction (G_{corr}) and Gibbs free energy (G) values at 298.15K for all the most stable configurations of the NA-SA-MA, NA-SA-DMA and NA-SA-A systems can be seen in Table S5.

Appendix A.2. Physical Principles of ACDC

In short, the core of ACDC is to use the birth–death equation (Equation (A1)) to describe the time-dependent cluster distributions:

$$\frac{dc_i}{dt} = \frac{1}{2}\sum_{j<i}\beta_{j,(i-j)}c_j c_{(i-j)} + \sum_j \gamma_{(i+j)\to i}c_{i+j} - \sum_j \beta_{i,j}c_i c_j - \frac{1}{2}\sum_{j<i}\gamma_{i\to j}c_i + Q_i - S_i \quad \text{(A1)}$$

where subscripts (i, j, $i - j$, and $i + j$) represent different clusters or monomers in the system. c_i represents the number concentration of i. $\beta_{i,j}$ denotes the collision rate coefficient between i and j, and $\gamma_{(i+j)\to i}$ denotes the evaporation rate of a cluster $i + j$ into smaller clusters, (or monomers) i and j. Q_i represents an additional outside source term of i, and S_i represents other sink terms for i. The collision rate coefficient ($\beta_{i,j}$) between clusters i and j is calculated using the hard-sphere kinetic gas theory as follows:

$$\beta_{i,j} = \left(\frac{3}{4\pi}\right)^{\frac{1}{6}}\left(\frac{6k_BT}{m_i} + \frac{6k_BT}{m_j}\right)^{\frac{1}{2}}\left(V_i^{\frac{1}{3}} + V_j^{\frac{1}{3}}\right)^2 \quad \text{(A2)}$$

where k_B is the Boltzmann constant, T is the temperature, and m_i and V_i are the mass and volume of i, respectively. The evaporation rate ($\gamma_{(i+j)\to i}$) is obtained from the corresponding collision rate coefficient and the binding free energy.

$$\gamma_{(i+j)\to i} = \beta_{i,j}c_{ref}\exp\left\{\frac{\Delta G_{i+j} - \Delta G_i - \Delta G_j}{k_BT}\right\} \quad \text{(A3)}$$

where ΔG is the binding free energy of the cluster and c_{ref} is the reference monomer concentration at 1 atm, which is the pressure at which ΔG was calculated.

Appendix A.3. Selection of Boundary Clusters

The boundary clusters fluxing out of the simulation box are required to be sufficiently stable (low evaporation rates). As shown in Figure 2, for the studied "3 × 3 box" NA-SA-MA system, binary $(SA)_3(MA)_{2\text{-}3}$ clusters and mixed ternary $(NA)_1(SA)_1(MA)_2$ and $(NA)_1(SA)_2(MA)_{2\text{-}3}$ and $(NA)_2(SA)_1(MA)_3$ clusters have relatively low evaporation rates on the order of 10^{-4} s^{-1}~10^{-2} s^{-1}. Therefore, $(SA)_4(MA)_3$, $(SA)_4(MA)_4$, $(NA)_1(SA)_3(MA)_3$, $(NA)_1(SA)_3(MA)_4$, $(NA)_2(SA)_2(MA)_4$ and $(NA)_3(SA)_1(MA)_4$ clusters were defined as boundary clusters in the ACDC simulation. For the NA-SA-DMA system, binary $(SA)_1(DMA)_1$, $(SA)_2(DMA)_{1\text{-}2}$ and $(SA)_3(DMA)_{2\text{-}3}$ clusters and mixed ternary $(NA)_1(SA)_1(DMA)_2$, $(NA)_1(SA)_2(DMA)_3$ and $(NA)_2(SA)_1(DMA)_3$ clusters have relatively low evaporation rates on the order of 10^{-5} s^{-1}~10^{-1} s^{-1}. Therefore, we selected $(SA)_4(DMA)_3$, $(SA)_4(DMA)_4$, $(NA)_1(SA)_3(DMA)_4$, $(NA)_2(SA)_2(DMA)_4$ and $(NA)_3(SA)_1(DMA)_4$ clusters as the boundary clusters in the ACDC simulation. For the NA-SA-A system, binary $(SA)_3(A)_{2\text{-}3}$ clusters and mixed ternary $(NA)_1(SA)_2(A)_3$ and $(NA)_2(SA)_1(A)_3$ clusters have relatively low evaporation rates on the order of 10^{-1} s^{-1}~10^1 s^{-1}. This allowed us to select $(SA)_4(A)_3$, $(SA)_4(A)_4$, $(NA)_1(SA)_3(A)_4$, $(NA)_2(SA)_2(A)_4$ and $(NA)_3(SA)_1(A)_4$ as the boundary clusters for the NA-SA-A system in the ACDC simulation.

References

1. Wang, M.; Kong, W.; Marten, R.; He, X.-C.; Chen, D.; Pfeifer, J.; Heitto, A.; Kontkanen, J.; Dada, L.; Kürten, A.; et al. Rapid Grov of New Atmospheric Particles by Nitric Acid and Ammonia Condensation. *Nature* **2020**, *581*, 184–189. [CrossRef] [PubMed]
2. Riccobono, F.; Schobesberger, S.; Scott, C.E.; Dommen, J.; Ortega, I.K.; Rondo, L.; Almeida, J.; Amorim, A.; Bianchi, F.; Brei lechner, M.; et al. Oxidation Products of Biogenic Emissions Contribute to Nucleation of Atmospheric Particles. *Science* **2(** *344*, 717–721. [CrossRef] [PubMed]
3. Yao, L.; Garmash, O.; Bianchi, F.; Zheng, J.; Yan, C.; Kontkanen, J.; Junninen, H.; Mazon, S.B.; Ehn, M.; Paasonen, P.; e Atmospheric New Particle Formation from Sulfuric Acid and Amines in A Chinese Megacity. *Science* **2018**, *361*, 278– [CrossRef] [PubMed]
4. Zhang, R.; Khalizov, A.; Wang, L.; Hu, M.; Xu, W. Nucleation and Growth of Nanoparticles in the Atmosphere. *Chem. Rev.* **2** *112*, 1957–2011. [CrossRef] [PubMed]
5. Zu, H.; Zhang, S.; Liu, L.; Zhang, X. The vital role of sulfuric acid in iodine oxoacids nucleation: Impacts of urban pollutants marine atmosphere. *Environ. Res. Lett.* **2024**, *19*, 014076. [CrossRef]
6. Kirkby, J.; Curtius, J.; Almeida, J.; Dunne, E.; Duplissy, J.; Ehrhart, S.; Franchin, A.; Gagné, S.; Ickes, L.; Kürten, A.; et al. Rol Sulphuric Acid, Ammonia and Galactic Cosmic Rays in Atmospheric Aerosol Nucleation. *Nature* **2011**, *476*, 429–433. [CrossF [PubMed]
7. Feketeová, L.; Bertier, P.; Salbaing, T.; Azuma, T.; Calvo, F.; Farizon, B.; Farizon, M.; Märk, T.D. Impact of A Hydrophobic Ion the Early Stage of Atmospheric Aerosol Formation. *Proc. Natl. Acad. Sci. USA* **2019**, *116*, 22540–22544. [CrossRef] [PubMed]
8. Chen, H.; Finlayson-Pitts, B.J. New Particle Formation from Methanesulfonic Acid and Amines/Ammonia as a Functior Temperature. *Environ. Sci. Technol.* **2017**, *51*, 243–252. [CrossRef] [PubMed]
9. Rose, C.; Zha, Q.; Dada, L.; Yan, C.; Lehtipalo, K.; Junninen, H.; Mazon, S.B.; Jokinen, T.; Sarnela, N.; Sipilä, M.; et al. Observati of Biogenic Ion-induced Cluster Formation in the Atmosphere. *Sci. Adv.* **2018**, *4*, eaar5218. [CrossRef]
10. Liu, L.; Zhong, J.; Vehkamäki, H.; Kurtén, T.; Du, L.; Zhang, X.; Francisco, J.S.; Zeng, X.C. Unexpected Quenching Effect on N Particle Formation from the Atmospheric Reaction of Methanol with SO_3. *Proc. Natl. Acad. Sci. USA* **2019**, *116*, 24966–24 [CrossRef]
11. Kirkby, J.; Duplissy, J.; Sengupta, K.; Frege, C.; Gordon, H.; Williamson, C.; Heinritzi, M.; Simon, M.; Yan, C.; Almeida, J.; e Ion-induced Nucleation of Pure Biogenic Particles. *Nature* **2016**, *533*, 521–526. [CrossRef] [PubMed]
12. Schobesberger, S.; Junninen, H.; Bianchi, F.; Lonn, G.; Ehn, M.; Lehtipalo, K.; Dommen, J.; Ehrhart, S.; Ortega, I.K.; Franchin, et al. Molecular Understanding of Atmospheric Particle Formation from Sulfuric Acid and Large Oxidized Organic Molecu *Proc. Natl. Acad. Sci. USA* **2013**, *110*, 17223–17228. [CrossRef] [PubMed]
13. Almeida, J.; Schobesberger, S.; Kurtén, A.; Ortega, I.K.; Kupiainen-Maatta, O.; Praplan, A.P.; Adamov, A.; Amorim, A.; Bianchi Breitenlechner, M.; et al. Molecular Understanding of Sulphuric Acid-amine Particle Nucleation in the Atmosphere. *Nature* **2(** *502*, 359–363. [CrossRef] [PubMed]
14. Kurtén, T.; Loukonen, V.; Vehkamäki, H.; Kulmala, M. Amines Are Likely to Enhance Neutral and Ion-induced Sulfuric Acid-wa Nucleation in the Atmosphere More Effectively than Ammonia. *Atmos. Chem. Phys.* **2008**, *8*, 4095–4103. [CrossRef]
15. Murphy, S.M.; Sorooshian, A.; Kroll, J.H.; Ng, N.L.; Chhabra, P.; Tong, C.; Surratt, J.D.; Knipping, E.; Flagan, R.C.; Seinf J.H. Secondary Aerosol Formation from Atmospheric Reactions of Aliphatic Amines. *Atmos. Chem. Phys.* **2007**, *7*, 2313–2 [CrossRef]
16. Smith, J.N.; Barsanti, K.C.; Friedli, H.R.; Ehn, M.; Kulmala, M.; Collins, D.R.; Scheckman, J.H.; Williams, B.J.; McMurry, I Observations of Aminium Salts in Atmospheric Nanoparticles and Possible Climatic Implications. *Proc. Natl. Acad. Sci. U* **2010**, *107*, 6634–6639. [CrossRef] [PubMed]
17. Loukonen, V.; Kurtén, T.; Ortega, I.K.; Vehkamäki, H.; Pádua, A.A.H.; Sellegri, K.; Kulmala, M. Enhancing Effect of Dimethylam in Sulfuric Acid Nucleation in the Presence of Water-A Computational Study. *Atmos. Chem. Phys.* **2010**, *10*, 4961–4974. [CrossR
18. Zhao, J.; Smith, J.N.; Eisele, F.L.; Chen, M.; Kuang, C.; McMurry, P.H. Observation of Neutral Sulfuric Acid-amine Contain Clusters in Laboratory and Ambient Measurements. *Atmos. Chem. Phys.* **2011**, *11*, 10823–10836. [CrossRef]
19. Lehtipalo, K.; Rondo, L.; Kontkanen, J.; Schobesberger, S.; Jokinen, T.; Sarnela, N.; Kuerten, A.; Ehrhart, S.; Franchin, A.; Niemir T.; et al. The Effect of Acid-base Clustering and Ions on the Growth of Atmospheric Nano-particles. *Nat. Commun.* **2016**, *7*, 11 [CrossRef]
20. Chen, M.; Titcombe, M.; Jiang, J.; Jen, C.; Kuang, C.; Fischer, M.L.; Eisele, F.L.; Siepmann, J.I.; Hanson, D.R.; Zhao, J.; e Acid-base Chemical Reaction Model for Nucleation Rates in the Polluted Atmospheric Boundary Layer. *Proc. Natl. Acad.* *USA* **2012**, *109*, 18713–18718. [CrossRef]
21. Erupe, M.E.; Viggiano, A.A.; Lee, S.H. The Effect of Trimethylamine on Atmospheric Nucleation Involving H_2SO_4. *Atmos. Ch* *Phys.* **2011**, *11*, 4767–4775. [CrossRef]
22. Lv, S.-S.; Miao, S.-K.; Ma, Y.; Zhang, M.-M.; Wen, Y.; Wang, C.-Y.; Zhu, Y.-P.; Huang, W. Properties and Atmospheric Implicat of Methylamine-Sulfuric Acid-Water Clusters. *J. Phys. Chem. A* **2015**, *119*, 8657–8666. [CrossRef] [PubMed]
23. Sipila, M.; Berndt, T.; Petaja, T.; Brus, D.; Vanhanen, J.; Stratmann, F.; Patokoski, J.; Mauldin, R.L.; Hyvarinen, A.P.; Lihavainen, et al. The Role of Sulfuric Acid in Atmospheric Nucleation. *Science* **2010**, *327*, 1243–1246. [CrossRef] [PubMed]

Olenius, T.; Halonen, R.; Kurtén, T.; Henschel, H.; Kupiainen-Määttä, O.; Ortega, I.K.; Jen, C.N.; Vehkamäki, H.; Riipinen, I. New Particle Formation from Sulfuric Acid and Amines: Comparison of Monomethylamine, Dimethylamine, and Trimethylamine. *J. Geophys. Res. Atmos.* **2017**, *122*, 7103–7118. [CrossRef]

Nadykto, A.B.; Yu, F.; Jakovleva, M.V.; Herb, J.; Xu, Y. Amines in the Earth's Atmosphere: A Density Functional Theory Study of the Thermochemistry of Pre-Nucleation Clusters. *Entropy* **2011**, *13*, 554–569. [CrossRef]

Nadykto, A.; Herb, J.; Yu, F.; Xu, Y.; Nazarenko, E. Estimating the Lower Limit of the Impact of Amines on Nucleation in the Earth's Atmosphere. *Entropy* **2015**, *17*, 2764–2780. [CrossRef]

Nadykto, A.B.; Herb, J.; Yu, F.; Xu, Y. Enhancement in the Production of Nucleating Clusters due to Dimethylamine and Large Uncertainties in the Thermochemistry of Amine-enhanced Nucleation. *Chem. Phys. Lett.* **2014**, *609*, 42–49. [CrossRef]

Jen, C.N.; McMurry, P.H.; Hanson, D.R. Stabilization of Sulfuric Acid Dimers by Ammonia, Methylamine, Dimethylamine, and Trimethylamine. *J. Geophys. Res. Atmos.* **2014**, *119*, 7502–7514. [CrossRef]

Qiu, C.; Zhang, R.Y. Multiphase Chemistry of Atmospheric Amines. *Phys. Chem. Chem. Phys.* **2013**, *15*, 5738–5752. [CrossRef]

Zhang, R.; Wang, L.; Khalizov, A.F.; Zhao, J.; Zheng, J.; McGraw, R.L.; Molina, L.T. Formation of Nanoparticles of Blue Haze Enhanced by Anthropogenic Pollution. *Proc. Natl. Acad. Sci. USA* **2009**, *106*, 17650–17654. [CrossRef]

Xu, W.; Zhang, R.Y. A Theoretical Study of Hydrated Molecular Clusters of Amines and Dicarboxylic Acids. *J. Chem. Phys.* **2013**, *139*, 064312. [CrossRef] [PubMed]

Myllys, N.; Elm, J.; Halonen, R.; Kurtén, T.; Vehkamäki, H. Coupled Cluster Evaluation of the Stability of Atmospheric Acid-Base Clusters with up to 10 Molecules. *J. Phys. Chem. A* **2016**, *120*, 621–630. [CrossRef] [PubMed]

Yu, H.; Ren, L.; Kanawade, V.P. New Particle Formation and Growth Mechanisms in Highly Polluted Environments. *Curr. Pollut. Rep.* **2017**, *3*, 245–253. [CrossRef]

Lee, S.H.; Gordon, H.; Yu, H.; Lehtipalo, K.; Haley, R.; Li, Y.; Zhang, R. New Particle Formation in the Atmosphere: From Molecular Clusters to Global Climate. *J. Geophys. Res. Atmos.* **2019**, *124*, 7098–7146. [CrossRef]

Cai, R.; Yan, C.; Yang, D.; Yin, R.; Lu, Y.; Deng, C.; Fu, Y.; Ruan, J.; Li, X.; Kontkanen, J.; et al. Sulfuric acid-Amine Nucleation in Urban Beijing. *Atmos. Chem. Phys.* **2021**, *21*, 2457–2468. [CrossRef]

Brown, S.S.; Ryerson, T.B.; Wollny, A.G.; Brock, C.A.; Peltier, R.; Sullivan, A.P.; Weber, R.J.; Dubé, W.P.; Trainer, M.; Meagher, J.F.; et al. Variability in Nocturnal Nitrogen Oxide Processing and Its Role in Regional Air Quality. *Science* **2006**, *311*, 67–70. [CrossRef] [PubMed]

Vayenas, D.V.; Takahama, S.; Davidson, C.I.; Pandis, S.N. Simulation of the Thermodynamics and Removal Processes in the Sulfate-Ammonia-Nitric acid system during Winter: Implications for $PM_{2.5}$ Control Strategies. *J Geophys. Res. Atmos.* **2005**, *110*, D07S14. [CrossRef]

Seinfeld, J.H.; Pandis, S.N. *Atmospheric Chemistry and Physics from Air Pollution to Climate Change*, 3rd ed.; Wiley: New York, NY, USA, 2016.

Kumar, M.; Zhong, J.; Zeng, X.C.; Francisco, J.S. Reaction of Criegee Intermediate with Nitric Acid at the Air-Water Interface. *J. Am. Chem. Soc.* **2018**, *140*, 4913–4921. [CrossRef]

Acker, K.; Möller, D.; Auel, R.; Wieprecht, W.; Kalaß, D. Concentrations of Nitrous acid, Nitric acid, Nitrite and Nitrate in the Gas and Aerosol Phase at a Site in the Emission Zone During ESCOMPTE 2001 Experiment. *Atmos. Res.* **2005**, *74*, 507–524. [CrossRef]

Mezuman, K.; Bauer, S.E.; Tsigaridis, K. Evaluating Secondary Inorganic Aerosols in Three Dimensions. *Atmos. Chem. Phys.* **2016**, *16*, 10651–10669. [CrossRef]

Song, C.H.; Carmichael, G.R. A Three-Dimensional Modeling Investigation of the Evolution Processes of Dust and Sea-Salt Particles in East Asia. *J. Geophys. Res. Atmos.* **2001**, *106*, 18131–18154. [CrossRef]

Zhang, R.; Sun, X.; Shi, A.; Huang, Y.; Yan, J.; Nie, T.; Yan, X.; Li, X. Secondary Inorganic Aerosols Formation During Haze Episodes at an Urban Site in Beijing, China. *Atmos. Environ.* **2018**, *177*, 275–282. [CrossRef]

Ding, J.; Zhao, P.; Su, J.; Dong, Q.; Du, X.; Zhang, Y. Aerosol pH and Its Driving Factors in Beijing. *Atmos. Chem. Phys.* **2019**, *19*, 7939–7954. [CrossRef]

Fairlie, T.D.; Jacob, D.J.; Dibb, J.E.; Alexander, B.; Avery, M.A.; van Donkelaar, A.; Zhang, L. Impact of Mineral Dust on Nitrate, Sulfate, and Ozone in Transpacific Asian Pollution Plumes. *Atmos. Chem. Phys.* **2010**, *10*, 3999–4012. [CrossRef]

Wang, M.; Xiao, M.; Bertozzi, B.; Marie, G.; Rörup, B.; Schulze, B.; Bardakov, R.; He, X.-C.; Shen, J.; Scholz, W. Synergistic HNO_3-H_2SO_4-NH_3 Upper Tropospheric Particle Formation. *Nature* **2022**, *605*, 483–489. [CrossRef]

Liu, L.; Li, H.; Zhang, H.; Zhong, J.; Bai, Y.; Ge, M.; Li, Z.; Chen, Y.; Zhang, X. The Role of Nitric Acid in Atmospheric New Particle Formation. *Phys. Chem. Chem. Phys.* **2018**, *20*, 17406–17414. [CrossRef]

Liu, L.; Yu, F.; Du, L.; Yang, Z.; Francisco, J.S.; Zhang, X. Rapid Sulfuric Acid-Dimethylamine Nucleation Enhanced by Nitric Acid in Polluted Regions. *Proc. Natl. Acad. Sci. USA* **2021**, *118*, e2108384118. [CrossRef]

Knattrup, Y.; Elm, J. Clusteromics IV: The Role of Nitric Acid in Atmospheric Cluster Formation. *ACS Omega* **2022**, *7*, 31551–31560. [CrossRef] [PubMed]

Knattrup, Y.; Kubečka, J.; Elm, J. Nitric Acid and Organic Acids Suppress the Role of Methanesulfonic Acid in Atmospheric New Particle Formation. *J. Phys. Chem. A* **2023**, *127*, 7568–7578. [CrossRef]

Wang, S.; Peng, Y.; Zhang, Q.; Wang, W.; Wang, Q. Mechanistic Understanding of Rapid H_2SO_4-HNO_3-NH_3 Nucleation in the Upper Troposphere. *Sci. Total. Environ.* **2023**, *883*, 163477. [CrossRef]

52. Schade, G.W.; Crutzen, P.J. Emission of Aliphatic Amines from Animal Husbandry and Their Reactions: Potential Source of N and HCN. *J. Atmos. Chem.* **1995**, *22*, 319–346. [CrossRef]
53. Xie, H.B.; Elm, J.; Halonen, R.; Myllys, N.; Kurten, T.; Kulmala, M.; Vehkamaki, H. Atmospheric Fate of Monoethanolami Enhancing New Particle Formation of Sulfuric Acid as an Important Removal Process. *Environ. Sci. Technol.* **2017**, *51*, 8422–8 [CrossRef] [PubMed]
54. Chen, H.; Varner, M.E.; Gerber, R.B.; Finlayson-Pitts, B.J. Reactions of Methanesulfonic Acid with Amines and Ammonia a Source of New Particles in Air. *J. Phys. Chem. B* **2016**, *120*, 1526–1536. [CrossRef] [PubMed]
55. Waller, S.E.; Yang, Y.; Castracane, E.; Racow, E.E.; Kreinbihl, J.J.; Nickson, K.A.; Johnson, C.J. The Interplay between Hydro Bonding and Coulombic Forces in Determining the Structure of Sulfuric Acid-Amine Clusters. *J. Phys. Chem. Lett.* **20** *9*, 1216–1222. [CrossRef] [PubMed]
56. Yang, Y.; Waller, S.E.; Kreinbihl, J.J.; Johnson, C.J. Direct Link between Structure and Hydration in Ammonium and Amini Bisulfate Clusters Implicated in Atmospheric New Particle Formation. *J. Phys. Chem. Lett.* **2018**, *9*, 5647–5652. [CrossRef]
57. McGrath, M.J.; Olenius, T.; Ortega, I.K.; Loukonen, V.; Paasonen, P.; Kurtén, T.; Kulmala, M.; Vehkamäki, H. Atmospheric Clus Dynamics Code: A Flexible Method for Solution of the Birth-death Equations. *Atmos. Chem. Phys.* **2012**, *12*, 2345–2355. [CrossR
58. Zhang, R.; Xie, H.B.; Ma, F.; Chen, J.; Iyer, S.; Simon, M.; Heinritzi, M.; Shen, J.; Tham, Y.J.; Kurten, T.; et al. Critical Role of Iod Acid in Neutral Iodine Oxoacid Nucleation. *Environ. Sci. Technol.* **2022**, *56*, 14166–14177. [CrossRef]
59. Cai, R.; Yin, R.; Li, X.; Xie, H.-B.; Yang, D.; Kerminen, V.-M.; Smith, J.N.; Ma, Y.; Hao, J.; Chen, J.; et al. Significant Contribution Trimethylamine to Sulfuric Acid Nucleation in Polluted Environments. *Npj Clim. Atmos. Sci.* **2023**, *6*, 75. [CrossRef]
60. Ma, F.; Xie, H.-B.; Zhang, R.; Su, L.; Jiang, Q.; Tang, W.; Chen, J.; Engsvang, M.; Elm, J.; He, X.-C. Enhancement of Atmosphe Nucleation Precursors on Iodic Acid-Induced Nucleation: Predictive Model and Mechanism. *Environ. Sci. Technol.* **20** *57*, 6944–6954. [CrossRef]
61. Elm, J. Elucidating the Limiting Steps in Sulfuric Acid-Base New Particle Formation. *J. Phys. Chem. A* **2017**, *121*, 8288–82 [CrossRef]
62. Zhang, J.; Dolg, M. ABCluster: The Artificial Bee Colony Algorithm for Cluster Global Optimization. *Phys. Chem. Chem. Ph* **2015**, *17*, 24173–24181. [CrossRef] [PubMed]
63. Zhang, J.; Dolg, M. Global Optimization of Clusters of Rigid Molecules using the Artificial Bee Colony Algorithm. *Phys. Ch Chem. Phys.* **2016**, *18*, 3003–3010. [CrossRef] [PubMed]
64. Frisch, M.J.; Trucks, G.W.; Schlegel, H.B.; Scuseria, G.E.; Robb, M.A.; Cheeseman, J.R.; Scalmani, G.; Barone, V.; Mennucci, Petersson, G.A.; et al. *Gaussian 09*; Gaussian, Inc.: Wallingford, CT, USA, 2009.
65. Neese, F. The ORCA program system. *Wiley Interdiscip. Rev. Comput. Mol. Sci.* **2012**, *2*, 73–78. [CrossRef]
66. Olenius, T.; Kupiainen-Määttä, O.; Ortega, I.K.; Kurtén, T.; Vehkamäki, H. Free Energy Barrier in the Growth of Sulfu Acid-Ammonia and Sulfuric Acid-Dimethylamine clusters. *J. Chem. Phys.* **2013**, *139*, 084312. [CrossRef] [PubMed]
67. Kupiainen-Määttä, O.; Olenius, T.; Kurtén, T.; Vehkamäki, H. CIMS Sulfuric Acid Detection Efficiency Enhanced by Amines D to Higher Dipole Moments: A Computational Study. *J. Phys. Chem. A* **2013**, *117*, 14109–14119. [CrossRef] [PubMed]
68. Temelso, B.; Morrison, E.F.; Speer, D.L.; Cao, B.C.; Appiah-Padi, N.; Kim, G.; Shields, G.C. Effect of Mixing Ammonia a Alkylamines on Sulfate Aerosol Formation. *J. Phys. Chem. A* **2018**, *122*, 1612–1622. [CrossRef] [PubMed]
69. Elm, J. Clusteromics I: Principles, Protocols, and Applications to Sulfuric Acid-Base Cluster Formation. *ACS Omega* **20** *6*, 7804–7814. [CrossRef]
70. Kubečka, J.; Neefjes, I.; Besel, V.; Qiao, F.; Xie, H.-B.; Elm, J. Atmospheric Sulfuric Acid-Multi-Base New Particle Format Revealed through Quantum Chemistry Enhanced by Machine Learning. *J. Phys. Chem. A* **2023**, *127*, 2091–2103. [CrossRef]

atmosphere

MDPI

icle

tegrated Analysis of Methane Cycles and Trends at the MO/GAW Station of Lamezia Terme (Calabria, Southern Italy)

ncesco D'Amico [1,2,*], Ivano Ammoscato [1], Daniel Gullì [1], Elenio Avolio [1], Teresa Lo Feudo [1], riafrancesca De Pino [1], Paolo Cristofanelli [3], Luana Malacaria [1], Domenico Parise [1], Salvatore Sinopoli [1], rgia De Benedetto [1] and Claudia Roberta Calidonna [1,*]

1 Institute of Atmospheric Sciences and Climate, National Research Council of Italy, Area Industriale Comp. 15, I-88046 Lamezia Terme, Catanzaro, Italy; i.ammoscato@isac.cnr.it (I.A.); d.gulli@isac.cnr.it (D.G); e.avolio@isac.cnr.it (E.A.); t.lofeudo@isac.cnr.it (T.L.F.); m.depino@isac.cnr.it (M.D.P.); l.malacaria@isac.cnr.it (L.M.); d.parise@isac.cnr.it (D.P.); s.sinopoli@isac.cnr.it (S.S.); g.debenedetto@isac.cnr.it (G.D.B.)

2 Department of Biology, Ecology and Earth Sciences, University of Calabria, Via Bucci Cubo 15B, I-87036 Rende, Cosenza, Italy

3 Institute of Atmospheric Sciences and Climate, National Research Council of Italy, Via P. Gobetti 101, I-40129 Bologna, Italy; p.cristofanelli@isac.cnr.it

* Correspondence: f.damico@isac.cnr.it or francesco.damico@unical.it (F.D.); claudiaroberta.calidonna@cnr.it (C.R.C.)

Abstract: Due to its high short-term global warming potential (GWP) compared to carbon dioxide, methane (CH_4) is a considerable agent of climate change. This research is aimed at analyzing data on methane gathered at the GAW (Global Atmosphere Watch) station of Lamezia Terme (Calabria, Southern Italy) spanning seven years of continuous measurements (2016–2022) and integrating the results with key meteorological data. Compared to previous studies on detected methane mole fractions at the same station, daily-to-yearly patterns have become more prominent thanks to the analysis of a much larger dataset. Overall, the yearly increase of methane at the Lamezia Terme station is in general agreement with global measurements by NOAA, though local peaks are present, and an increase linked to COVID-19 is identified. Seasonal changes and trends have proved to be fully cyclic, with the daily cycles being largely driven by local wind circulation patterns and synoptic features. Outbreak events have been statistically evaluated depending on their weekday of occurrence to test possible correlations with anthropogenic activities. A cross analysis between methane peaks and specific wind directions has also proved that local sources may be deemed responsible for the highest mole fractions.

Keywords: methane; wind parameters; GAW; Lamezia Terme; Calabria; Southern Italy

1. Introduction

Methane (CH_4) is the simplest among alkanes and, due to its capacity to absorb the infrared radiation in a band at the 1.7, 2.3, 3.3, and 7.6 μm [1], is a very potent greenhouse gas (GHG), with a GWP (global warming potential) 84–87 times higher than that of CO_2 for the time span of two decades [2]. Methane's GWP-100 value (GWP in a century) drops to 27 compared to CO_2 [3]; as a GHG, it is much less persistent in the atmosphere compared to carbon dioxide, thus falling in the short-lived climate forcers (SLCF) category, along with ozone and aerosols [2]. Present-day atmospheric concentrations of methane are higher compared to pre-industrial levels: this is well documented in the literature ever since the late 1990s [4]. Atmospheric CH_4 mole fractions have recently surpassed the 150% increase threshold compared to pre-industrial era levels [5]. In fact, the current estimate for 1750 levels is approximately 700 ppb, three times lower than the global mean of 1911.9 ppb observed in 2022 [6]. The increase of methane in the atmosphere is influenced by anthropogenic activities [7,8]. On a global scale, CH_4 has been measured by NOAA's Earth System Research Laboratory ever since 1983, so the data from that point onward are deemed particularly reliable in the effort of defining short-to-long atmospheric trends [9]. Methane has also proved to drive climate change via several side effects to the chemistry of Earth's atmosphere, such as the stratospheric production of H_2O and the release of tropospheric O_3 (ozone) [2]. Although methane concentrations are two orders of magnitude lower compared to those of carbon dioxide, the high GWP of this compound has sparked notable interest in the scientific community and is now widely regarded as one of the main causes of climate change [10–12]. Overall, methane contributes to 19% of the global radiative forcing from the pre-industrial era to 2022, compared to carbon dioxide's forcing of 64% [13].

Estimates on present-day methane releases, uptake, and sinks have several degrees of uncertainty [14], but improvements in predictive models over the past few years have helped to somehow constrain these estimates. Globally, according to the IEA report from 2021 [15], the annual emissions of methane—both natural and anthropogenic—are approximately 570 Tg (teragrams, 10^{12} g). It is worth noting that some estimates are as high as 737 Tg [12]. Geologic releases account for 43–50 Tg y^{-1} [16], while the broad anthropogenic emissions are in the 360 Tg y^{-1} range, 110–128 Tg of which are related to pure fossil fuel burning [12]. Wetlands contribute with 101–179 Tg y^{-1} [12], biomass burning estimates are as high as 30 Tg y^{-1} [12], livestock releases account for an estimated 95–109 Tg y^{-1}, 87–97 Tg of which are directly linked to enteric fermentation processes [17]. Termites are deemed responsible for 15 Tg y^{-1} worth of emissions [18]. In the past few years, the nature and characteristics of geologic emissions have been further divided and classified into sub-categories, depending on their sources and production mechanisms (e.g., geothermal manifestations, microseepage, macro-seeps) [16]. These categories are now fully recognized by IPCC reports. Overall, it is now estimated that 40% of methane emissions are natural, while the remaining 60% are anthropogenic [12].

Sinks and uptake phenomena remove approximately 630 Tg of methane from the atmosphere each year [10], but the amount is variable over time [12]. The geographical distribution of methane emissions is asymmetric, with the northern hemisphere yielding higher values compared to the southern hemisphere; seasonal cycles have also been observed even before the age of NOAA's Earth System Research Laboratory-enhanced detections [19]. Additionally, soil uptake of methane in tropical and temperate forests of the southern hemisphere is higher than that of their northern counterparts [20]. The overall upward trend in atmospheric concentration is attributable to the excess in anthropogenic emissions compared to natural sinks [10], though the phenomenon of methane variability in Earth's atmosphere is much more complicated compared to that of CO_2: for instance, an anomalous drop in atmospheric methane was recorded in 2004 [21], with the whole 1983–2006 observation streak reporting a downward trend in annual growth rates [22–24]. Methane has been on a nearly constant rise ever since that sudden drop, but the isotopic $^{13}C/^{12}C$ ratio ($\delta^{13}C$) began to decrease after two centuries of regular increase, highlighting a major shift in fractionation processes induced by various sinks, as well as other mechanisms [5,25]. Isotopic ratios aside, a well above average increase was recorded in 2020, attributable to the COVID-19 pandemic and other factors, such as changes in the hydroxyl radical (OH) sink (see Section 5) and increased natural emissions from wetlands [26]. Overall, while the trends and mechanisms that drive CO_2 increases in the atmosphere are very well defined, those of CH_4 could benefit from additional data availability [27], hence the need of more detailed analyses.

This work provides unprecedented detail on methane cycles and trends detected by the World Meteorological Organization/Global Atmosphere Watch (WMO/GAW) Lamezia Terme (code: LMT) observation site in Calabria (Southern Italy) over a period of seven years, from 2016 to 2022. This study is meant to be a step forward in the monoparameter characterization (in this case, aimed specifically at CH_4) of an observation site in Europe, following examples such as the work of Padilla et al. (2023) [28]. The second section of this manuscript will describe the observation site and its characteristics, while the third section will report information on datasets and methods used to process data. The results are shown in the fourth section, followed by discussion, conclusions, and perspectives on future research studies. The Supplementary Materials cover graphs and tables not shown in the main text.

2. The Lamezia Terme CNR-ISAC Observatory

The Lamezia Terme site (WMO/GAW code: LMT) is a coastal station located in Calabria, Southern Italy, more precisely, in the Sant'Eufemia plain (Lat: 38.83° N; Lon: 16.23° E; Alt: 6 m above sea level), south of Lamezia Terme. The observatory, fully operated by the National Research Council of Italy-Institute of Atmospheric Sciences and Climate (CNR-ISAC), is located approximately 600 m from the Tyrrhenian coastline of Calabria, and officially started its data gathering operations in 2015. It has since provided continuous data on several chemical and meteorological parameters. Wind characteristics at this coastal site were analyzed in the past [29]. Due to the geographic location of the experimental site, winds coming from the northeast are more subject to anthropogenic influence while winds coming from the west, largely influenced by the sea, yield lower concentrations in pollutants, as clearly demonstrated in an earlier study [30]. The prevailing W-NE local wind circulation is well established in sectors other than atmospheric research: in fact, the magnetic bearing of the runway in use at the nearby Lamezia Terme International Airport (IATA: SUF; ICAO: LICA) is 10/28 (100–280° N).

Ever since it started operations, LMT reported methane concentrations that may be attributable to several nearby sources, hence the concept of a "multisource" coastal site. Cristofanelli et al. (2017) [30], accounting for one year of data gathering, indicated the nearby Lamezia Terme International Airport (located 3 km north from the observation site) and local livestock farming as two possible explanations for the observed peaks in methane, in addition to the anthropogenic pollution from nearby urban areas that is channeled towards the observatory when winds are coming from the northeast. as well as the A2 highway (part of the European route E45) located nearby (Figure 1B). In addition to the above-mentioned sources, it is worth noting the presence of *Lamezia Terme Centrale* (Lamezia Terme Central Station), which is a hub for railway connections between the main Tyrrhenian/cross country line and the Ionian line on the opposite coast of the region. The train station is located 5.4 km N-NE of LMT. Two landfills are also located nearby, and their locations are marked in Figure 1B. Farms are spread over the area.

From a climatological point of view, the area is known to be affected by seasonal changes to wind circulation. The study area, due to its geographical location, is mainly affected by breeze circulation, as assessed by several works.

Federico et al. (2010a) [29], in studying the characteristics and importance of breezes, demonstrated how they dominate the local circulation and play a major role in the local climate. The authors observed relevant changes in average wind speeds over the seasons, as well as slight changes in wind orientation, though the general W-WSW/NE-ENE axis is dominant throughout the whole year and is deemed the result of wind channeling through the Marcellinara gap, which is the narrowest point in the entire Italian peninsula. When the 850 hPa layer is considered, the preferred orientation switches to a prevailing NW direction, which is consistent with large scale circulation in the area. Federico et al. (2010b) [31] further analyzed the nature of breeze circulations, considering two years of observation and modeling analyses, concluding that during spring, summer, and part of fall, diurnal breeze circulation is considered as a combination of local and large-scale flows, while large-scale forcing represents a main driver of diurnal circulation in winter, November included, and nocturnal flows are believed to be due to the circulation of nocturnal breezes.

The confirmation on the presence of breeze and the influence of local conditions were both confirmed by further studies such as Gullì et al. (2017) [32] via the analysis of wind–lidar profiles for two continuous years. Figure 2 shows wind profiles accounting for multiple altitude thresholds, from 10 up to 300 m, collected by a Zephir Lidar 300 at the LMT observation site.

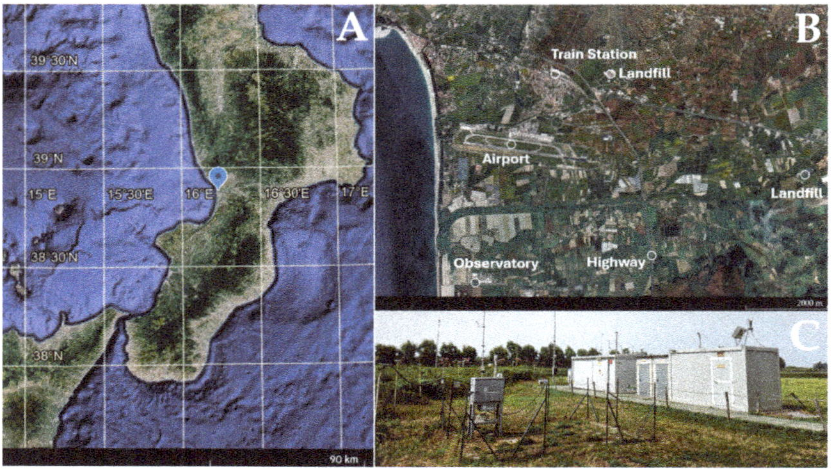

Figure 1. (**A**): The location of Lamezia Terme's observation site within the region of Calabria, Southern Italy. (**B**): Details of the southwestern Lamezia Terme municipal area, with marks on the location of the observatory and key infrastructures/locations addressed in this research study, namely, SUF/LICA airport, the central train station, and two landfills. The "Highway" mark refers to a specific point to the east of LMT where the distance between the A2 highway itself and the observatory is approximately 4.3 km. Farms, including those with livestock, are spread over the area. (**C**): Details of the LMT observation site, where all instruments are located.

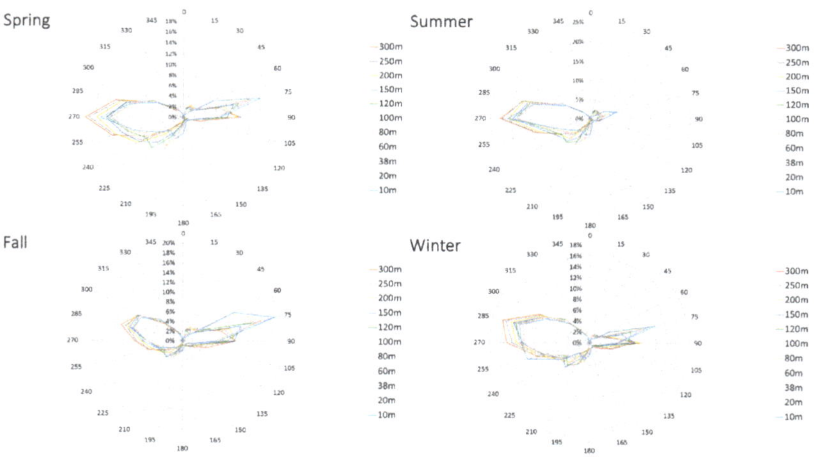

Figure 2. Seasonal wind roses at all levels during the 2014–2016 period, divided by thresholds between 10 and 300 m, and collected by Zephir Wind Lidar. From Gullì et al. (2017) [32], modified.

In a recent work, Lo Feudo et al. (2020) [33] studied the characteristics of the vertical structure of the Planetary Boundary Layer (PBL) at the LMT site during an experimental campaign carried out in July/August 2009. The integration of different instruments (surface stations, wind profiler, lidar, and sodar) with high resolution weather model products allows researchers to further assess the roles played by sea breeze and synoptic flows in the area.

Due to its location in the articulated context of the Mediterranean and, specifically, in the narrowest corridor of the Italian peninsula, the LMT station is subject to exposure to open fire emissions, as recently reported in Malacaria et al. (2024) [34], and frequent Saharan dust events [35].

3. Instruments, Datasets, and Methods

Data on methane (ppb) mole fractions have been gathered by a Picarro G2401 CRDS (Cavity Ring-Down Spectrometry) analyzer (Santa Clara, CA, USA). This high-precision instrument measures gas concentrations in the atmospheric medium thanks to their capacity to absorb and scatter light at specific wavelengths, which ensures high-precision estimates on mole fractions. The same instrument also gathers data on CO (carbon monoxide, in ppm), CO_2 (carbon dioxide, in ppm), and H_2O (water vapor, as a percentage). Specifically, the G2401 CRDS analyzer operates via a four-point configuration: one gathers ambient air, while the other three are connected to three standard calibration cylinders produced by the WMO/GAW Central Calibration Laboratory hosted by NOAA (National Oceanic and Atmospheric Administration) GML (Global Monitoring Laboratory) to ensure comparability with the WMO X2004 mole fraction scale. The reference cylinders cover CH_4 fractions in the 300–2600 nmol/mol range. A Vici-Valco rotative valve (model: EMTMA-CE) switches between the four configurations at regular intervals. In particular, calibration cylinders are measured every 14 days. Each cylinder is measured for a time span of 30 min during three measurement cycles (360 min). Furthermore, three target cylinders are measured for quality check purposes every 19 h. A Nafion dryer (model: MD-070-144S-4) reduces the impact of water vapor on measurements by drying ambient air. Dried air is then analyzed by the G2401. Measurements occur every 5 s with a precision of 1 ppb. For the purpose of this research, hourly averages have been considered and their respective standard deviations are reported throughout this article.

An automatic weather station (Vaisala WXT520, Vantaa, Finland) measured at 10 m ASL the following meteorological parameters: temperature, relative humidity, wind speed and direction, pressure, and rain (10 averaged minutes). For the purposes of this research, wind speed and wind directions have been considered and are measured by the WXT520 instrument via ultrasound. Wind speed data have an uncertainty of ± 0.3 m/s, while wind direction data have an uncertainty of ± 3 sexagesimal degrees. Wind data, just like in the previous case, have been aggregated on an hourly basis featuring the standard deviation as a data stability indicator.

Table 1 shows the coverage of available methane data compared to the total amount of hours elapsed between 1 January 2016 and 31 December 2022 (61,368). Overall, of the 61,368 elapsed hours, 57,990 (94.49%) have been covered by verified and calibrated methane detections. This dataset will be referred to as "primary" throughout the paper. Three years out of seven have a total coverage rate exceeding 95%. Please note that both 2016 and 2020 are leap years, with an extra 24 h each (8784 h instead of 8760).

Table 1 also reports the overall coverage of Picarro G2401 and Vaisala WXT520 data compared to the actual number of hours of the entire observation period. Data analysis involving both chemical data on methane and meteorological wind data was based on a subset of the main database where both instruments were fully operating at the same time. This, by definition, led to a slightly smaller "secondary" dataset available for cross-analyses of this particular kind. Five out of seven years have a combined integrated data coverage of more than 90%.

Finally, the two datasets have been processed in R 4.4.0 via ggplot2, ggpubr, tidyverse, and openair packages, as well as their respective libraries. Depending on the purpose of each analysis, data have also been aggregated on a seasonal basis (JFD = December, January, February for Winter; MAM = March, April, May for Spring; JJA = June, July, August for Summer; SON = September, October, November for Fall). Yearly, monthly, and daily aggregations have also been computed.

Table 1. Main coverage parameters (expressed as percentage of total actual hours per year) of the entire dataset spanning seven years of continuous observations at LMT are shown in this table. Also reported is the coverage rate per year, with a maximum total coverage of 99.57% in 2017, and a minimum coverage of 83.83% in 2022. The second column shows the integration of Picarro G2401 and Vaisala WXT520 data satisfying the condition where both instruments were reliably operating at the same time.

Year	Picarro Coverage (%)	Picarro-Vaisala (%)
2016	94.92%	92.2%
2017	99.57%	93.37%
2018	94%	74.68%
2019	97.6%	97.57%
2020	93.8%	93.79%
2021	97.71%	97.46%
2022	83.83%	75.41%
	94.49% [1]	89.07% [1]

[1] Total coverage.

4. Results

4.1. General Monthly Trend

The longer dataset compared to previous research [30] highlights more long-term trends and cycles. Figure 3 shows monthly aggregated averages where seasonal trends, as well as a general upward trend, can be noticed. The 1st and 3rd quartiles are also plotted in Figure 3. Higher values are generally linked to Winter and Fall seasons, while Summer and Spring seasons yield the lowest values per year. A November 2017 peak can be noticed in pre-pandemic years and will be discussed in Section 5.

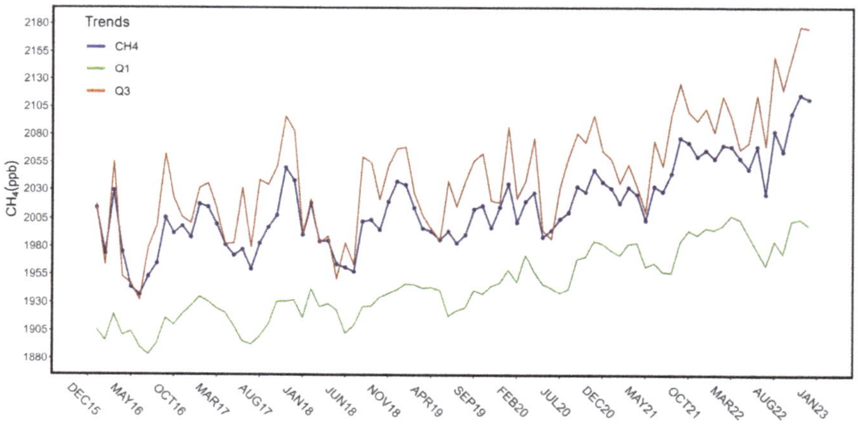

Figure 3. Monthly averages accounting for the entire 2016–2022 period (84 months), shown in dark blue. Attenuated brown and green lines show 3rd and 1st quartile trends, respectively.

4.2. Daily Cycle

With respect to daily cycles, intended as variations over the course of 24 h, Figure 4 shows the hourly variation of methane mole fractions. The figure shows seasonal 2019 data specifically, as it is the year with the highest degree of integrated methane and wind data coverage (97.57%, see Table 1). Data are from the primary dataset. Table 2 reports the average values per hour, divided by season, as well as the standard deviations computed during data evaluation. Data and graphs concerning years other than 2019 are available as Supplementary Materials S1(A1) through S1(F3): the y axis of these graphs is set to the same scale to allow direct comparisons between years. Figure 4(A2,B2) have a different scale compared to A1,B1 meant to visualize the daily cycle both in terms of absolute concentrations (A2) and standard deviations (B2).

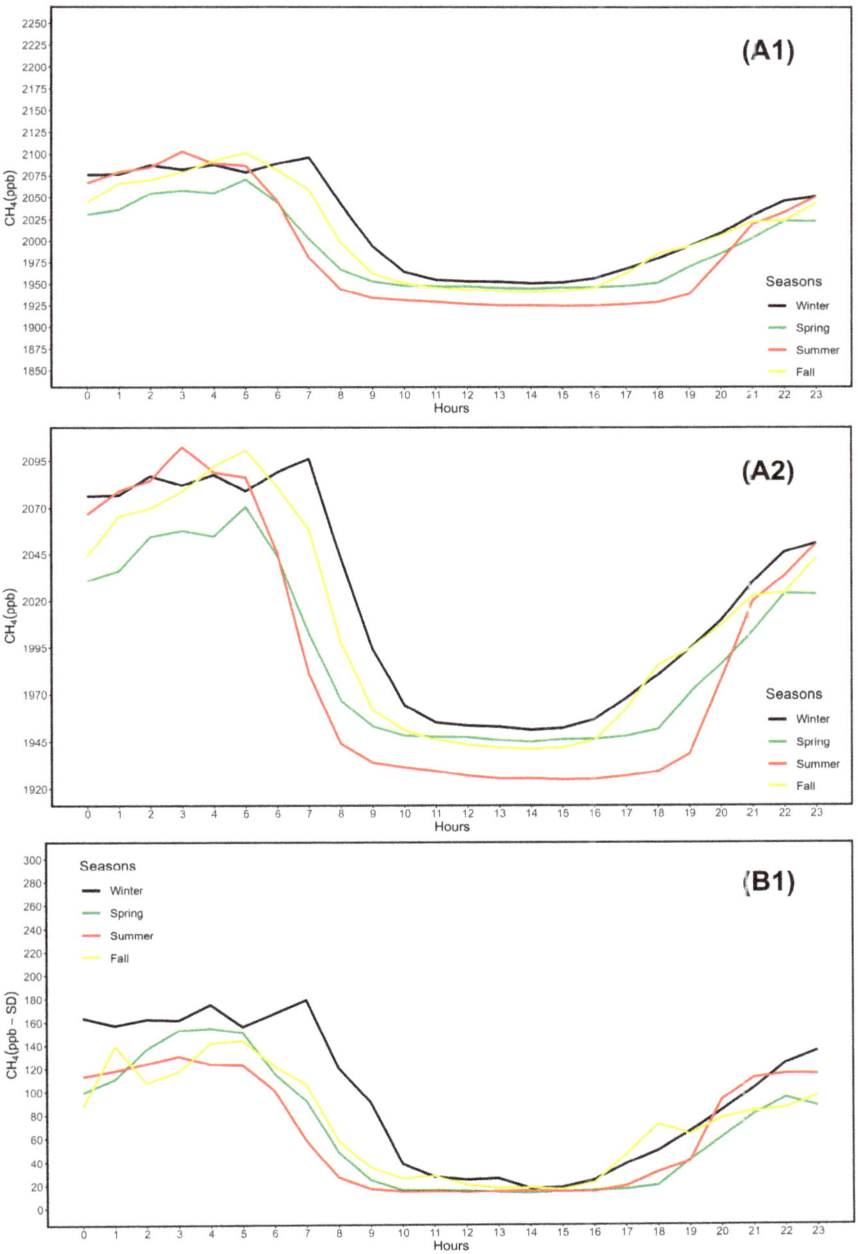

Figure 4. *Cont.*

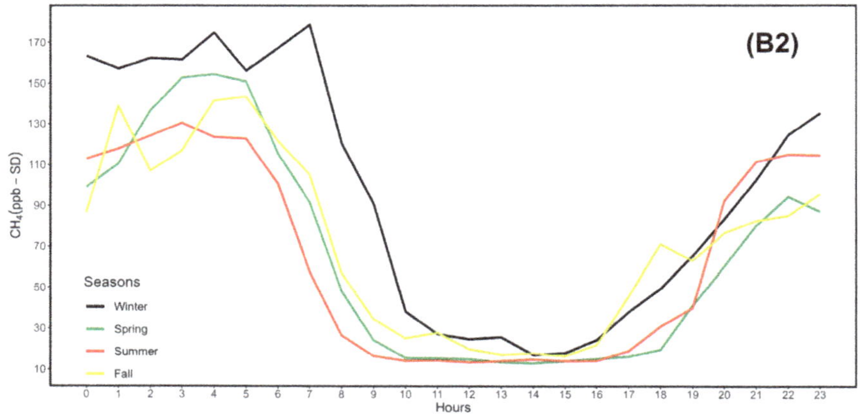

Figure 4. Hourly averages of methane concentrations, showing seasonal variations. The color scheme is the following: Winter (black), Spring (green), Summer (red), Fall (yellow). Please note that, as stated above, the graph is referring to 2019 data only. (**A1**): methane mole fractions; (**A2**): same data, with the y axis range narrowed down to highlight hourly changes; (**B1**): methane standard deviations; (**B2**): narrowed down graph.

Table 2. The hourly values plotted in Figure 4 are listed, with the addition of standard deviations for each seasonal parameter. The table refers specifically to 2019.

Hours	Winter	Win. SD	Spring	Spr. SD	Summer	Sum. SD	Fall	Fa. SD
0	2076.27	163.49	2030.78	99.30	2066.69	113.04	2043.76	86.57
1	2076.78	157.14	2036.03	110.72	2078.96	118.00	2065.42	138.91
2	2087.11	162.39	2054.75	136.81	2084.67	124.21	2069.45	107.13
3	2082.18	161.63	2057.92	152.80	2102.78	130.50	2078.47	116.90
4	2087.75	174.86	2054.98	154.43	2088.88	123.60	2091.81	141.44
5	2079.08	156.14	2070.65	150.94	2086.37	122.84	2100.68	143.55
6	2089.18	167.36	2043.86	115.58	2045.54	100.81	2080.75	121.62
7	2096.24	178.82	2002.39	91.74	1980.23	57.72	2058.33	105.30
8	2043.12	120.53	1966.65	48.25	1943.77	26.59	1997.36	57.00
9	1994.23	90.90	1953.08	24.33	1933.72	16.59	1961.31	34.57
10	1964.34	38.20	1948.24	15.71	1931.18	14.27	1950.67	25.13
11	1955.17	27.22	1947.67	15.55	1929.20	14.54	1945.88	27.87
12	1953.52	24.69	1947.42	15.17	1926.67	13.62	1943.12	19.89
13	1952.92	25.71	1945.74	13.58	1925.17	14.20	1941.48	17.17
14	1951.27	17.15	1944.92	13.12	1925.21	15.02	1940.85	17.86
15	1952.23	18.01	1946.33	14.14	1924.65	14.22	1941.62	16.52
16	1956.75	24.34	1946.45	15.31	1924.85	14.36	1945.54	21.89
17	1967.86	38.03	1947.92	16.32	1926.58	19.07	1961.88	45.35
18	1979.94	49.42	1951.55	19.72	1928.99	31.05	1985.03	71.06
19	1994.36	65.39	1970.26	41.22	1938.43	40.10	1994.08	62.93
20	2009.53	83.48	1985.83	60.23	1977.44	92.46	2006.87	76.47
21	2030.08	102.69	2003.80	80.21	2019.90	111.61	2022.63	82.51
22	2046.53	124.62	2024.07	94.40	2033.38	115.17	2024.07	85.05
23	2051.48	135.37	2023.57	87.10	2051.06	114.79	2042.95	95.66

Overall, a prominent daily cycle is reported in Figure 4. Seasonal differences can also be noticed, with the Winter and Summer seasons yielding the highest and lowest values, respectively, though this does not occur throughout the entire observation period. The "flat valley" in both absolute methane concentrations and hourly standard deviations occurring between 10:00 and 16:00 UTC is no coincidence, as the influence of local wind circulation on observed concentrations is known to be substantial. This interval coincides, in fact, with prevalent SW-NW sectors. At 03:00 UTC, the standard deviations are in the 116.9–161.63 ppb range (Fall and Winter, respectively), while at 15:00 UTC, the range is 14.14–18.01 ppb. Identical graphs and tables, though applied to the other years, are accessible as Supplementary Materials S1A through S1G. Graphs showing the variations in standard deviations are also accessible from these materials.

4.3. Wind Data Evaluation

Pollution roses referring to 2019 are shown in Figure 5, grouped by season. For these evaluations, the secondary dataset described in Table 1 was used. As reported in Section 2, the observation site is affected by two distinct wind circulation corridors: a pure western-seaside sector yielding generally lower methane concentrations, and a northeastern-continental sector showing much higher concentrations. Pollution roses reporting data referring to the other years are accessible as Supplementary Materials S2A through S2F.

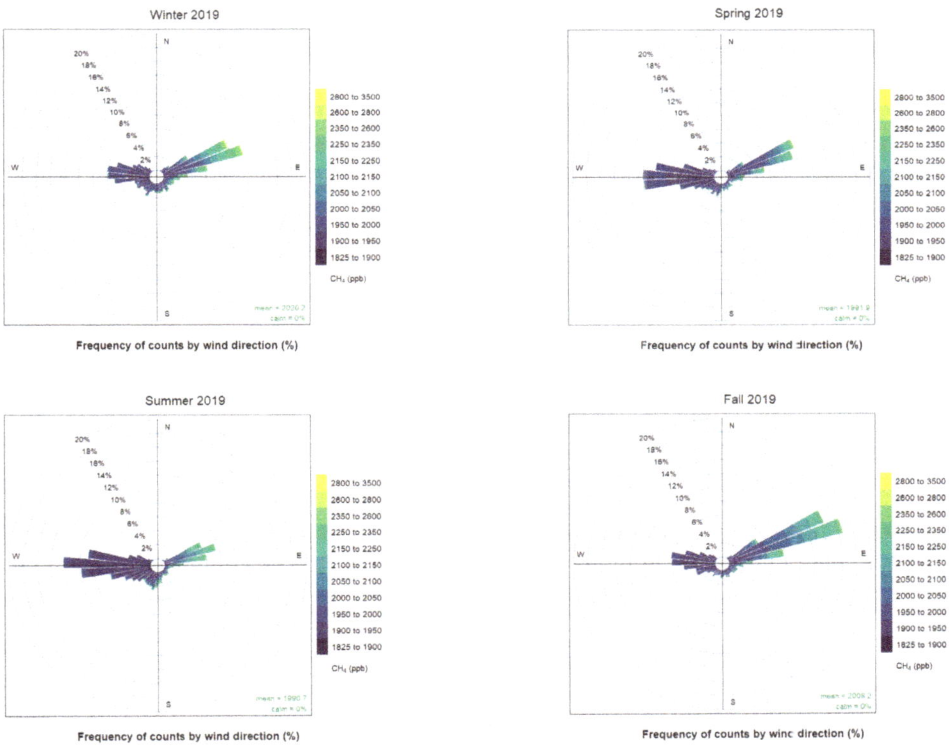

Figure 5. Seasonal pollution roses of methane concentrations observed at LMT and referring to 2019. Each bar represents an angle of 8 degrees.

Via the application of two distinct filters, observed methane concentrations have been correlated with wind speeds. In particular, a western-seaside sector (240–300° N, 3260 h) and a northeastern-continental (0–90° N, 2921 h) sector have been filtered from the secondary dataset, constituting, respectively, 38.13% and 34.17% of 2019's available data. An amount of 27.7% of detections fall outside these ranges. The results are shown in Figure 6 and clearly indicate how the highest methane concentrations are linked to northeastern winds, in conjunction with lower wind speeds (Figure 6B). Winds coming from the west sector yield considerably lower methane concentrations, even though wind speeds are variable and winter-time peaks exceeding 14 m/s can be noticed (Figure 6A). Also, note how—if combined—the two sectors would lean to a hyperbola branch distribution, as shown in Supplementary Material S3G (the supplementary graph also includes values falling outside the two ranges). Supplementary Materials S3A through S3F show the same graphs applied to years other than 2019, while S3G shows all observed values.

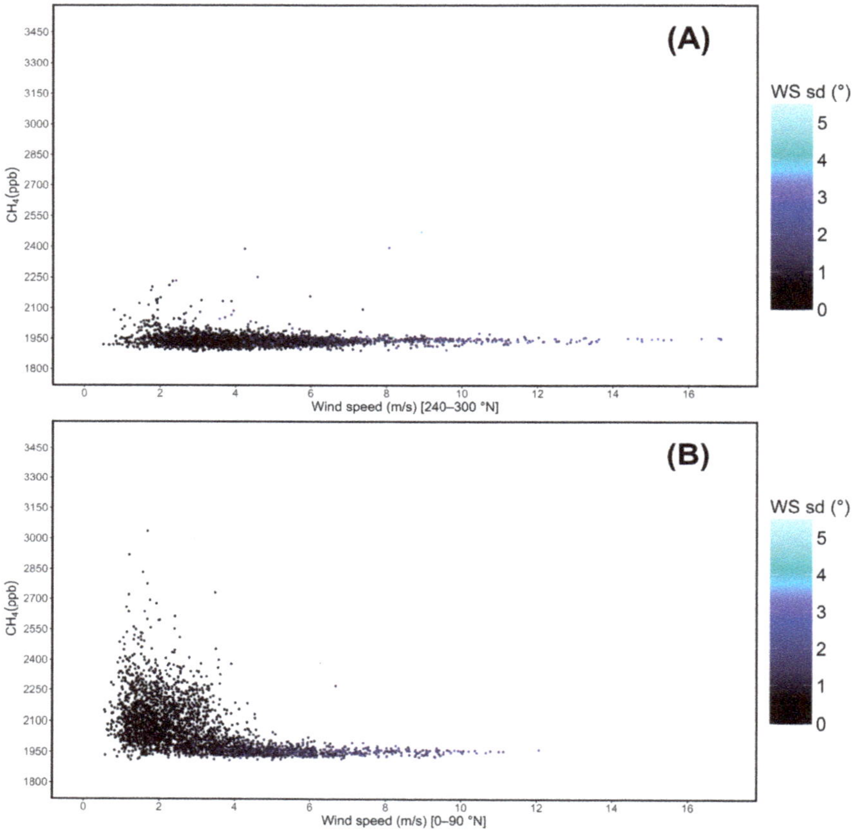

Figure 6. Correlations between observed wind speeds (*X* axis) and methane concentrations (*Y* axis). (**A**): Western-seaside sector (240–300° N). (**B**): Northeastern-continental sector (0–90° N). The color scheme is set to represent hourly changes in wind speed standard deviations.

4.4. Outbreak Analysis

A further evaluation was aimed specifically at possible indicators of anthropogenic activities, which could be susceptible to weekdays, under the assumption that no natural mechanism would lead to substantial statistical differences affecting the occurrence of outbreak events during a week. In the case of 2019, two distinct evaluations have been performed on a per-weekday basis, one accounting for values equal to or greater than the 3rd quartile (2029.11 ppb), and a second, more constrained evaluation accounting for the top 2.5% values (2299.91 ppb). The analysis was performed on the primary Picarro G2401 dataset. Respectively, 2138 and 214 h satisfied these conditions, now plotted in Figure 7. Supplementary Materials S4-A through S4-F show results concerning years other than 2019.

Table 3 shows the two filters, including the number of hours satisfying their conditions, throughout the entire 2016–2022 period.

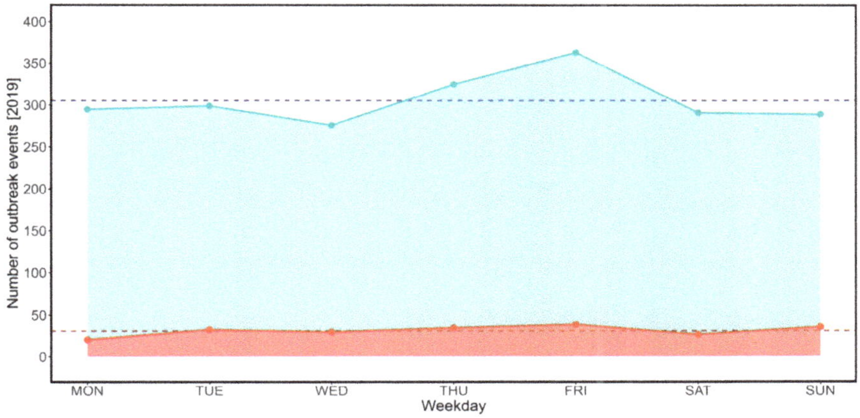

Figure 7. Weekly (MON–SUN) distribution of outbreak events exceeding the 3rd quartile (turquoise) and 97.5% threshold (red) in 2019. The two y axis intercepts show the average number of outbreak events per weekday, which is 305.71 and 30.57, respectively. Shaded areas are meant to better discriminate the two categories.

Table 3. The 3rd quartile and 97.5% threshold counts data and details, per year. The last row shows the total number of hours exceeding the two limits, as well as the average count of hours satisfying that condition, by weekday.

Year	3rd Q. (ppb)	Hours ≥ 3rd Q.	Average Count Per Weekday (3rd Q.)	97.5% Threshold (ppb)	Hours ≥ 97.5% Threshold	Average Count per Weekday (97.5%)
2016	1994.2	2085	297.85	2418.76	209	29.85
2017	2031.23	2181	311.57	2401.73	219	31.28
2018	2017.44	2059	294.14	2346.05	206	29.42
2019	2029.11	2138	305.71	2299.91	214	30.57
2020	2056.46	2060	294.28	2278.84	206	29.42
2021	2069.5	2140	305.71	2337.82	214	30.57
2022	2117.58	1836	262.28	2432.91	184	26.28
		14,499 [1]	2071.28 [2]		1452 [1]	207.42 [2]

[1] Total amount. [2] Average value.

A Chi-squared test was performed on combined data concerning outbreak events listed by weekday. Said test was executed in R 4.4.0 by setting a value of 9999 Monte Carlo replicates. Table 4 reports the data used, as well as the computed χ^2 and p-values for both the 3rd quartile and 97.5% threshold. For the graph, see Supplementary Material S4G. Though over the course of the observation period, a variability in peaks per weekday can be noticed (see Supplementary Materials S4A through S4F), combined 2016–2022 data point to Friday as the weekday with the most frequent occurrence of outbreaks. Due to its higher χ^2 and lower p-value, the 3rd quartile category has provided a more statistically relevant result in terms of distribution.

Table 4. Results of the Chi-squared test performed to verify the possible outbreak occurrence susceptibility to specific weekdays. For the graph, see Supplementary Material S4G.

Type	Average	MON	TUE	WED	THU	FRI	SAT	SUN	χ^2	*p*-Valu
3rd Q.	2071.28	1899	2024	2089	2163	2246	2082	1996	37.152	0.0001
97.5% th.	207.42	181	179	210	201	251	207	223	17.817	0.0064

4.5. Multi-Year Trend

Finally, a multi-year trend has been plotted in Figure 8. For this evaluation, the year 2022 has been excluded due to its lower coverage rate of 83.83% compared to the other years, which fall in the 93.8–99.57% range (see Table 1).

Table 5. Annual means of methane concentrations and their respective standard deviations, in ppb. The year 2022, written in italics, is excluded from Figures 7 and 8. NOAA annual means and changes are extrapolated from Lan et al. (2024) [36].

Year	CH4 (ppb)	CH4 SD (ppb)	Coverage (%)	Change (ppb)	NOAA (ppb)	NOAA Change (ppb)	LMT-NOA Diff. (ppb
2016	1980.87	146.78	94.92	-	1843.12	+7.05	137.75
2017	1999.75	140.30	99.57	+18.88	1849.67	+6.89	150.08
2018	1994.88	121.38	94	−4.87	1857.33	+8.70	137.55
2019	2002.50	106.98	97.6	+7.62	1866.58	+9.67	135.92
2020	2018.79	96.20	93.8	+16.29	1878.93	+15.17	139.86
2021	2040.79	105.99	97.71	+22.00	1895.28	+17.91	145.51
2022 [1]	*2074.87*	*126.87*	*83.83*	*+34.07*	*1922.53*	*+13.26*	*137.75*

[1] 2022 is reported in italics but not featured in Figures 8 and 9.

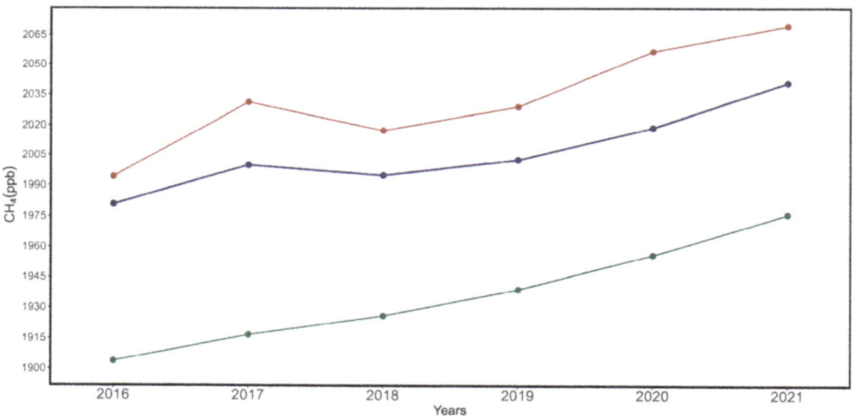

Figure 8. Average annual methane values over the course of LMT's observation history (dark blue), except for 2022 due to its lower coverage rate compared to 2016–2021 (see Tables 1 and 5). Also included are annual trends in the 3rd quartile (dark red) and 1st quartile (dark green) values.

The global annual means issued by NOAA and LMT's observations have been compared. Differences between annual means are shown in Table 5. Overall, with the exception of a surge in 2017 (difference: 150.08 ppb), five annual differences in the 2016–2021 period fall in the 135.92–139.86 ppb range, with the 2022 divergence being identical to that of 2016. Both trends are shown in Figure 9.

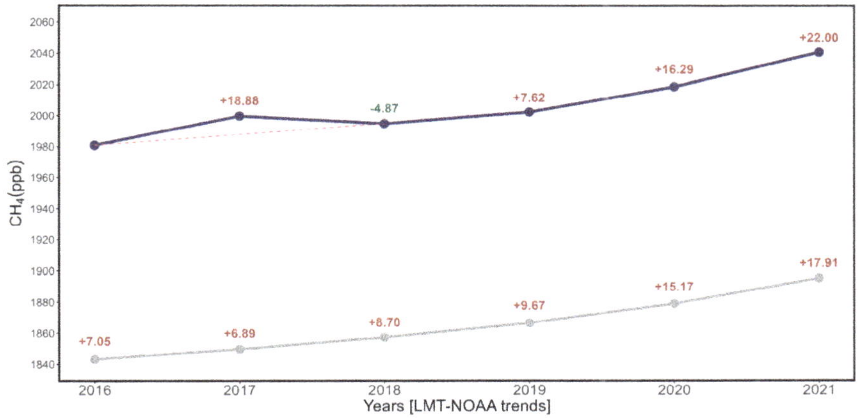

Figure 9. Direct comparison of LMT (dark blue) and global NOAA (dark gray) annual means. Annual variations are also reported. The local 2017 surge at LMT is highlighted via a dashed line connecting 2016 and 2018 annual averages.

Both Table 5 and Figure 9 remark on the 2020 surges, which are likely related to the COVID-19 pandemic, which is known to have led to an increase in global CH_4 concentrations [36,37], though other factors need to be considered to justify these surges (see Section 5). On average, LMT annual means exceed the global NOAA values by 141.11 ppb. The 2017 surge is local, and further information on 2017 trends can be obtained via Supplementary Materials S1(B1), S1(B2), S1(B3) (daily cycles), S2B (seasonal pollution roses), and S3(B1), S3(B2) (wind speed and mole fractions).

5. Discussion

The larger dataset and the integration of key meteorological information have allowed for a more detailed analysis on methane concentrations detected at the LMT observatory. The detailed analysis on methane detections from the 0–90° N range allows researchers to better constrain some of the hypotheses made in the past on local sources of pollution and integrate them with new hypotheses. Research such as Cristofanelli et al. (2017) [30] proposed local livestock farming, air traffic, domestic heating (Winter-specific), and landfills among the causes of higher GHG and reactive gas concentrations in the area. In the case of methane, this is compatible with observed wind directions at LMT, as the locations of at least one landfill, the A2 highway, nearby urban areas, and specific take-off/landing trajectories at SUF/LICA airport, are indeed compatible with these wind regimes, methane peaks, and their respective seasonal variations.

The general trend seen in Figure 3 highlights seasonal cycles and an upward trend towards higher concentrations over time, while the daily cycles seen in Figure 4 are dominated by wind circulation. "Flat valleys" in hourly graphs are linked to western winds. The further analysis of methane concentrations with respect to wind directions (Figure 5) and speed (Figure 6) has allowed us to determine the existence of two main corridors: a western-seaside direction, linked to lower methane concentrations regardless of wind speed, and a northeastern-continental direction, which is characterized by higher methane mole fractions (especially in the case of low wind speeds), which are likely due to nearby sources of this compound. The corridors are susceptible to seasonal variations, as reported in Figure 5 and related Supplementary Materials.

For the first time in the data-gathering history of the observation site LMT, a possible correlation between weekdays and outbreak events was tested under the assumption that no mechanism in nature other than anthropogenic factors would lead to statistically significant differences in outbreak event occurrence over the course of weekdays. Figure 7 and

Table 4, as well as related Supplementary Materials S4A through S4G, provide statistical relevance to higher results on Fridays, though it is worth noting that each year's results show several degrees of variability. Friday peaks may be the result of additional anthropogenic emissions related to commuting, public transportation, farming, and industrial activities, though future works will have to investigate these occurrences even further. A previous study on black carbon (BC), $PM_{2.5}$, and PM_{10} concentrations already reported higher weekday/weekend ratios as an indicator of additional anthropogenic activity [38].

Overall, the results shown in this work with respect to methane concentration variability depending on wind direction, combined with the statistical distribution of outbreaks over the course of weekdays, seem to corroborate the hypothesis seen in previous studies such as Cristofanelli et al. (2017) [30] by which anthropogenic sources located in the northeast with respect to LMT are responsible for generally higher methane concentrations in the area. In the context of a "multisource" scenario, the previous study reported, in particular, two possible sources of methane in the area: local livestock farming and air traffic. A case can be made, following the research by Malacaria et al. (2024) [34], that wildfires also constitute methane sources affecting LMT detections, but this needs to be corroborated by future research, possibly accounting for additional atmospheric tracers. These findings could then be used by regulators and policy makers to address issues related to local air quality concerns.

Livestock such as cattle are indeed responsible for major methane emission levels worldwide [39,40] and contribute up to an estimated 14.5% of total anthropogenic greenhouse gas emissions. Specifically, Hristov et al. (2013) [41] reported that each cow releases 60–160 kg CH_4 y^{-1}, while goats and sheep release 10–16 kg CH_4 y^{-1}, all depending on characteristics such as the dry matter intake parameter (DMI) and size of the ruminant. In the case of observed methane concentrations at LMT, it is currently impossible to pinpoint emissions strictly related to livestock farming in the area unless new parameters such as carbon isotope fractionation are considered. This extra parameter will discriminate livestock methane from fossil fuel output and other sources of this compound.

Similarly, the influence of local air traffic is also difficult to estimate for a number of reasons, including operational characteristics and aeronautical procedures. One preliminary conclusion would be that the observatory and the airport's runways intersect two parallel air corridors which do not affect each other, but the leading literature on air traffic pollution and emissions place the LMT station inside the "near-airport" (<10 km) range category, which is—according to studies such as Carslaw et al. (2006) [42] and Carslaw and Beevers, (2013) [43]—subject to direct air traffic influence. Aircraft engine combustion processes are among the known sources of methane and do contribute to anthropogenic climate change [44,45]. The literature on air quality perturbation by air traffic has also focused on LTO (landing to take-off) cycles, which are related to operational phases during which aircraft are either on the ground or at low altitudes [46,47], while research on broader effects of air traffic also considers cruise phases [48,49]. The airport was recognized in previous research as a possible influence over LMT's detections [30], but no source apportionment has been performed. Specifically, runway (RWY) 10 take-offs (aircraft facing east) and RWY 28 landings (aircraft facing west) would place aircraft on trajectories compatible with the northeastern-continental wind sector shown in Figure 5 and related Supplementary Materials. It is presently not possible, without the introduction of additional tracers, to provide tangible estimates concerning the airport's influence over local methane detections.

Finally, relevant assumptions can be made when considering multi-year trends observed at LMT. Globally, methane concentrations are on the rise as a result of anthropogenic activities, natural emissions (e.g., wetlands), and major variations in sinks [26,36]. On a local scale—though the clear upward trend persists—sporadic "bursts" of concentrations may occur, such as the 2017 peak reported in Figures 8, 9, and Table 5. In the case of LMT, November 2017 yields a peak average value of 2049.74 $\pm$ 178.2 ppb (mean $\pm$ 1σ), which does not have a clear explanation. Supplementary Materials S1(B1), S1(B2), S1(B3) (daily cycles), S2B (seasonal pollution roses), and S3(B1), S3(B2) (wind speed and mole fractions), all referring to 2017, point to higher concentrations from the NE sector, that we attributed mainly to anthropogenic activities. However, it is not possible to pinpoint specific emission sectors linked to these peaks; we do exclude wildfires as a possible source due to the season involved (Fall). In 2020, a year heavily affected by the COVID-19 pandemic, global values experienced a 15 ppb increase [14,36], which is similar to the 16.29 ppb surge observed at LMT. It is worth noting that an annual global increase of 1 ppb is believed to be the result of an extra $\approx$2.77 Tg CH_4 being emitted into the atmosphere [5]. The 2019–2020 local-to-global leap is not surprising, as the COVID-19 outbreak lockdowns are among the causes of increases in methane concentrations [37]. Methane has a latency of approximately one decade in the atmosphere, but several models clearly show that the peak response of this alkane occurs within a few months from extra-emission pulses [50,51], corroborating the hypothesis by which the 2020 peak is at least partially linked to nation-wide lockdowns and consequent increases in emissions from the energy sector [52], though the 2020 peak is likely due to a combination of multiple factors. A recent study by Feng et al. (2023) [53] estimated that 66% of the 2020 methane surge was due to increased emission rates, though it is worth noting that changes in sinks caused by reduced NO_x have also been reported as leading causes for such surges [26]. Also, although on the global scale an upward trend is detected, findings at the LMT observation site confirm concentration levels that are also dependent on latitude and hemisphere [19,20], which would explain why LMT yearly trends follow a pattern similar to that of global trends observed by NOAA (Figure 9), with a difference in terms of absolute concentrations that is also driven by its geographical context and fluxes occurring at local and regional scales.

Moreover, it is worth specifying that CH_4 reacts with the hydroxyl radical (OH), which is the main compound responsible for methane sink in natural environments [54,55]. The hydroxyl is not directly measured at LMT, so direct comparisons and correlations between the trends of both compounds are not possible. It is recognized, however, that the availability and diffusion of hydroxyl is a major controlling factor for methane and is, therefore, responsible for net atmospheric increases and losses [7], though several uncertainties in its trends and quantifications exist [56]. Exceptional events such as the VEI-6 eruption of Mount Pinatubo in 1991 and the consequent release of SO_2 and sulfates in the atmosphere have reportedly reduced the sink of CH_4 from reaction with OH in the troposphere by approximately 17.8 Tg between June 1991 and June 1993 [57]. Chlorine also contributes to active methane removal from the atmosphere and estimates such as those proposed by Hossaini et al. (2016) [58] provide an approximate value of 12–13 Tg y^{-1}, though further research indicates that this sink may be less than half than that [59]. Despite its minor role, the chlorine sink has a strong isotopic fingerprint which can be determined via adequate equipment and instruments [60,61]. Future research at LMT, accounting for the carbon isotope fractionation of methane, could potentially provide information on the influence of the chlorine sink. Furthermore, by contrast to methanogenesis induced by microbial activity [62], methanotroph bacteria in water and soils also contribute to the removal of methane [63,64], but as soils are constantly threatened by human activities [65], forest soils' methane uptake capacity is also compromised, especially considering the fact that it is already vulnerable to environmental factors such as precipitation [66]. There is, however, evidence going in the opposite direction, with consistent reports of increasing forest soil methane uptake [20]. Trends aside, these drivers of methane removal help reduce to nearly one decade the latency of this compound in the atmosphere, while carbon dioxide can persist for several centuries, at most, one millennium [3]. At least in the case of the hydroxyl radical, it has been demonstrated that post-2006 methane atmospheric increases cannot be solely attributed to a decrease in atmospheric OH [14], thus adding up even more to the general complexity of methane buildup into the atmosphere. The complexity behind the interaction of methane with the hydroxyl radical, therefore, requires more research be aimed at the investigation of sinks and their effects observed on CH_4 mole fractions.

6. Conclusions

For the first time, multi-year trends of methane detected by the Lamezia Terme WMO/GAW station (LMT) have been analyzed. Located close to the Tyrrhenian coast of Calabria, in the narrowest point of the Italian peninsula, measurements at this observatory are largely influenced by its peculiar location in the country. Daily cycles are influenced by local wind circulation and synoptic features: a western-seaside sector yields lower methane concentrations, while northeastern-continental winds yield the highest concentrations detected at LMT. The circulation is such that daytime winds come mostly from the western sector, thus resulting in much lower methane values, while nocturnal winds come from the northeast. Filtering data by wind direction clearly demonstrates the differences in terms of average concentrations as detected from the two sectors. Moreover, wind speed is also very closely tied to methane values, as the comparison of methane concentrations and wind speeds results in a hyperbola branch pattern where low speeds are linked to higher values and, vice versa, high speeds yielded lower values: this is interpreted as the influence of local emissions outputs. Seasonal differences and cycles are also well defined, with Winter and Summer seasons generally yielding the highest and lowest methane concentrations, respectively. Multi-year trends have been compared with NOAA's global measurements: the upward tendencies are similar and differences between yearly averages fall mostly within a well-defined window (137.75–145.51 ppb; 2017 has a peak of 150.08 ppb), though it is worth noting that methane concentrations are known from the literature to vary depending on geographical factors such as latitude. For the first time, a method was introduced to correlate outbreak events of methane concentrations with weekdays in the effort to determine possible anthropogenic influences over these values, as these activities alone do have a weekly cycle which is totally lacking in nature. The results indicate that peaks tend to occur on Fridays, while Mondays and Sundays yielded lower values. The introduction of $\delta^{13}C$ measurements of CH_4, as well as CO_2, would significantly help with source apportionment in future studies.

Supplementary Materials: The following supporting information can be downloaded at https://www.mdpi.com/article/10.3390/atmos15080946/s1, Figure S1-A1, -B1, -C1, -D1, -E1, -F1: daily cycles of methane concentrations shown on a seasonal basis (2016–2018; 2020–2022); Figure S1-A2, -B2, -C2, -D2, -E2, -F2: daily cycles in methane standard deviations, shown on a seasonal basis (2016–2018; 2020–2022); Table S1-A3, -B3, -C3, -D3, -E3, -F3: daily methane mole fractions and standard deviations referred to 2016–2018 and 2020–2022. Figure S2-A, -B, -C, -D, -E, -F. Seasonal pollution roses of methane concentrations, referred to 2016-2018 and 2020–2022; Figure S3-A1, -B1, -C1, -D1, -E1, -F1: methane mole fractions and wind speeds from the western (240–300° N) sector, referred to 2016–2018 and 2020–2022; Figure S3-A2, -B2, -C2, -D2, -E2, -F2: methane mole fractions and wind speeds from the northeastern (0–90° N) sector, referred to 2016–2018 and 2020–2022; Figure S3-G: methane mole fractions and wind speeds accounting for all wind directions throughout the entire 2016–2022 observation period; Figure S4-A, -B, -C, -D, -E, -F: weekly distribution of outbreak events (third quartile and top 2.5% interval), referred to 2016–2018 and 2020–2022; Figure S4-G: weekly distribution of outbreak events accounting for the entire 2016–2022 observation period.

Author Contributions: Conceptualization, F.D. and C.R.C.; methodology, F.D., C.R.C. and T.L.F.; software, F.D.; validation, C.R.C., T.L.F. and P.C.; formal analysis, F.D.; investigation, F.D.; data curation, F.D., I.A., D.G., E.A., T.L.F., P.C., L.M., D.P., S.S. and G.D.B.; writing—original draft preparation, F.D.; writing—review and editing, F.D., C.R.C., I.A., D.G., E.A., T.L.F., M.D.P., P.C., L.M., D.P., S.S. and G.D.B.; visualization, F.D., C.R.C., D.G. and T.L.F.; supervision, C.R.C. and P.C.; funding acquisition, C.R.C. and M.D.P. All authors have read and agreed to the published version of the manuscript.

Funding: This research was funded by AIR0000032-ITINERIS, the Italian Integrated Environmental Research Infrastructures System (D.D. n. 130/2022-CUP B53C22002150006) under the EU-Next Generation EU PNRR-Mission 4 "Education and Research"-Component 2: "From research to business"-Investment 3.1: "Fund for the realization of an integrated system of research and innovation infrastructures".

Institutional Review Board Statement: Not applicable.

Informed Consent Statement: Not applicable.

Data Availability Statement: The datasets presented in this article are not readily available because they are part of other ongoing studies. However, tables shown in the main text and Supplementary Materials are accessible, providing that their source is cited.

Acknowledgments: The authors would like to thank the support and assistance of the editor from Atmosphere editorial office. They would also like to thank the three anonymous reviewers who contributed to expand and improve the manuscript.

Conflicts of Interest: The authors declare no conflicts of interest.

erences

Byrom, R.E.; Shine, K.P. Methane's Solar Radiative Forcing. *Geophys. Res. Lett.* **2022**, *49*, e2022GL098270. [CrossRef]
Myhre, G.; Shindell, D.; Bréon, F.M.; Collins, W.; Fuglestvedt, J.; Huang, J.; Koch, D.; Lamarque, J.F.; Lee, D.; Mendoza, B.; et al. Anthropogenic and Natural Radiative Forcing. In *Climate Change 2013: The Physical Science Basis, Contribution of Working Group I to the Fifth Assessment Report of the Intergovernmental Panel on Climate Change*; Intergovernmental Panel on Climate Change: Cambridge, UK; New York, NY, USA, 2013.
Sand, M.; Skeie, R.B.; Sandstad, M.; Krishnan, S.; Myhre, G.; Bryant, H.; Derwent, R.; Hauglustaine, D.; Paulot, F.; Prather, M.; et al. A multi-model assessment of the Global Warming Potential of hydrogen. *Nat. Commun.* **2023**, *4*, 203. [CrossRef]
Etheridge, D.M.; Steele, L.P.; Francey, R.J.; Langenfelds, R.L. Atmospheric methane between 1000 A.D. and present: Evidence of anthropogenic emissions and climatic variability. *J. Geophys. Res. Atmos.* **1998**, *103*, 15979–15993. [CrossRef]
Nisbet, E.G.; Manning, M.R.; Dlugokencky, E.J.; Fisher, R.E.; Lowry, D.; Michel, S.E.; Lund Myhre, C.; Platt, S.M.; Allen, G.; Bousquet, P.; et al. Very Strong Atmospheric Methane Growth in the 4 Years 2014–2017: Implications for the Paris Agreement. *Glob. Biogeochem. Cycles* **2019**, *33*, 318–342. [CrossRef]
Blunden, J.; Boyer, T.; Bartow-Gillies, E. State of the Climate in 2022. *B. Am. Meteorol. Soc.* **2023**, *104*, 1–501. [CrossRef]
Skeie, R.B.; Hodnebrog, Ø.; Myhre, G. Trends in atmospheric methane concentrations since 1990 were driven and modified by anthropogenic emissions. *Commun. Earth Environ.* **2023**, *4*, 317. [CrossRef]
Szopa, S.; Naik, V.; Adhikary, B.; Artaxo, P.; Berntsen, T.; Collins, W.D.; Fuzzi, S.; Gallardo, L.; Kiendler-Scharr, A.; Klimont, Z.; et al. Short-Lived Climate Forcers. In *Climate Change 2021: The Physical Science Basis. Contribution of Working Group I to the Sixth Assessment Report of the Intergovernmental Panel on Climate Change*; Masson-Delmotte, V., Zhai, P., Pirani, A., Connors, S.L., Péan, C., Berger, S., Caud, N., Chen, Y., Goldfarb, L., Gomis, M.I., et al., Eds.; Cambridge University Press: Cambridge, UK; New York, NY, USA, 2021; pp. 817–922.
Dlugokencky, E.J.; Steele, L.P.; Lang, P.M.; Masarie, K.A. The growth rate and distribution of atmospheric methane. *J. Geophys. Res. Atmos.* **1994**, *99*, 17021–17043. [CrossRef]
Saunois, M.; Bousquet, P.; Poulter, B.; Peregon, A.; Ciais, P.; Canadell, J.G.; Dlugokencky, E.J.; Etiope, G.; Bastviken, D.; Houweling, S.; et al. The global methane budget 2000–2012. *Earth Syst. Sci. Data* **2016**, *8*, 697–751. [CrossRef]
Saunois, M.; Bousquet, P.; Poulter, B.; Peregon, A.; Ciais, P.; Canadell, J.G.; Dlugokencky, E.J.; Etiope, G.; Bastviken, D.; Houweling, S.; et al. Variability and quasi-decadal changes in the methane budget over the period 2000–2012. *Atmos. Chem. Phys.* **2017**, *17*, 11135–11161. [CrossRef]
Saunois, M.; Stavert, A.R.; Poulter, B.; Bousquet, P.; Canadell, J.G.; Jackson, R.B.; Raymond, P.A.; Dlugokencky, E.J.; Houweling, S.; Patra, P.K.; et al. The Global Methane Budget 2000–2017. *Earth Syst. Sci. Data* **2020**, *12*, 1561–1623. [CrossRef]
Montzka, S.A. The NOAA annual greenhouse gas index (AGGI). *NOAA Glob. Monit. Lab.* **2023**. Available online: https://gml.noaa.gov/aggi/aggi.html (accessed on 16 July 2024).
Lan, X.; Nisbet, E.G.; Dlugokencky, E.J.; Michel, S.E. What do we know about the global methane budget? Results from four decades of atmospheric CH_4 observations and the way forward. *Philos. Trans. A Math. Phys. Eng. Sci.* **2021**, *379*, 20200440. [CrossRef] [PubMed]
IEA Methane Tracker 2021. International Energy Agency, Paris, 2021. Available online: https://www.iea.org/reports/methane-tracker-2021 (accessed on 16 July 2024).
Etiope, G.; Ciotoli, G.; Schwietzke, S.; Schoell, M. Gridded maps of geological methane emissions and their isotopic signature. *Earth Syst. Sci. Data* **2019**, *11*, 1–22. [CrossRef]
Chang, J.; Peng, S.; Ciais, P.; Saunois, M.; Dangal, S.R.S.; Herrero, M.; Havlík, P.; Tian, H.; Bousquet, P. Revisiting enteric methane emissions from domestic ruminants and their $\delta^{13}C_{CH4}$ source signature. *Nat. Commun.* **2019**, *10*, 3420. [CrossRef] [PubMed]
Dlugokencky, E.; Nisbet, E.G.; Fisher, R.; Lowry, D. Global atmospheric methane: Budget, changes and dangers. *Philos. Trans. A Math. Phys. Eng. Sci.* **2011**, *369*, 2058–2072. [CrossRef] [PubMed]
Blake, D.R.; Mayer, E.W.; Tyler, S.C.; Makide, Y.; Montague, D.C.; Rowland, F.S. Global increase in atmospheric methane concentrations between 1978 and 1980. *Geophys. Res. Lett.* **1982**, *9*, 477–480. [CrossRef]
Yu, L.; Huang, Y.; Zhang, W.; Li, T.; Sun, W. Methane uptake in global forest and grassland soils from 1981 to 2010. *Sci. Total Environ.* **2017**, *607–608*, 1163–1172. [CrossRef] [PubMed]

21. Pison, I.; Ringeval, B.; Bousquet, P.; Prigent, C.; Papa, F. Stable atmospheric methane in the 2000s: Key-role of emissions fr natural wetlands. *Atmos. Chem. Phys.* **2013**, *13*, 11609–11623. [CrossRef]
22. Dlugokencky, E.; Masarie, K.; Lang, P.; Tans, P.P. Continuing decline in the growth rate of the atmospheric methane bur *Nature* **1998**, *393*, 447–450. [CrossRef]
23. Dlugokencky, E.J.; Houweling, S.; Bruhwiler, L.; Masarie, K.A.; Lang, P.M.; Miller, J.B.; Tans, P.P. Atmospheric methane levels Temporary pause or a new steady-state? *Geophys. Res. Lett.* **2003**, *30*, 1992. [CrossRef]
24. He, J.; Naik, V.; Horowitz, L.W.; Dlugokencky, E.; Thoning, K. Investigation of the global methane budget over 1980–2017 us GFDL-AM4.1. *Atmos. Chem. Phys.* **2020**, *20*, 805–827. [CrossRef]
25. Ferretti, D.F.; Miller, J.B.; White, J.W.C.; Etheridge, D.M.; Lassey, K.R.; Lowe, D.C.; Macfarling Meure, C.M.; Dreier, M.F.; Trudin C.M.; Van Ommen, T.D.; et al. Unexpected changes to the global methane budget over the past 2000 years. *Science* **2005**, 1714–1717. [CrossRef] [PubMed]
26. Peng, S.; Lin, X.; Thompson, R.L.; Xi, Y.; Liu, G.; Hauglustaine, D.; Lan, X.; Poulter, B.; Ramonet, M.; Saunois, M.; et al. Wetl emission and atmospheric sink changes explain methane growth in 2020. *Nature* **2022**, *612*, 477–482. [CrossRef]
27. Wilson, C. Untangling variations in the global methane budget. *Nat. Commun.* **2023**, *4*, 318. [CrossRef]
28. Padilla, R.; Adame, J.A.; Hidalgo, P.J.; Bolivar, J.P.; Yela, M. Short-term trend and temporal variations in atmospheric methan an Atlantic coastal site in Southwestern Europe. *Atmos. Environ.* **2023**, *333*, 120665. [CrossRef]
29. Federico, S.; Pasqualoni, L.; De Leo, L.; Bellecci, C. A study of the breeze circulation during summer and fall 2008 in Calab Italy. *Atmos. Res.* **2010**, *97*, 1–13. [CrossRef]
30. Cristofanelli, P.; Busetto, M.; Calzolari, F.; Ammoscato, I.; Gullì, D.; Dinoi, A.; Calidonna, C.R.; Contini, D.; Sferlazzo, D.; Di Io T.; et al. Investigation of reactive gases and methane variability in the coastal boundary layer of the central Mediterranean ba *Elem. Sci. Anth.* **2017**, *5*, 12. [CrossRef]
31. Federico, S.; Pasqualoni, L.; Sempreviva, A.M.; De Leo, L.; Avolio, E.; Calidonna, C.R.; Bellecci, C. The seasonal characteristic the breeze circulation at a coastal Mediterranean site in South Italy. *Adv. Sci. Res.* **2010**, *4*, 47–56. [CrossRef]
32. Gullì, D.; Avolio, E.; Calidonna, C.R.; Lo Feudo, T.; Torcasio, R.C.; Sempreviva, A.M. Two years of wind-lidar measurements a Italian Mediterranean Coastal Site. In European Geosciences Union General Assembly 2017, EGU—Division Energy, Resource Environment, ERE. *Energy Procedia* **2017**, *125*, 214–220. [CrossRef]
33. Lo Feudo, T.; Calidonna, C.R.; Avolio, E.; Sempreviva, A.M. Study of the Vertical Structure of the Coastal Boundary La Integrating Surface Measurements and Ground-Based Remote Sensing. *Sensors* **2020**, *20*, 6516. [CrossRef]
34. Malacaria, L.; Parise, D.; Lo Feudo, T.; Avolio, E.; Ammoscato, I.; Gullì, D.; Sinopoli, S.; Cristofanelli, P.; De Pino, M.; D'Amico et al. Multiparameter detection of summer open fire emissions: The case study of GAW regional observatory of Lamezia Ter (Southern Italy). *Fire* **2024**, *7*, 198. [CrossRef]
35. Calidonna, C.R.; Avolio, E.; Gullì, D.; Ammoscato, I.; De Pino, M.; Donateo, A.; Lo Feudo, T. Five Years of Dust Episode the Southern Italy GAW Regional Coastal Mediterranean Observatory: Multisensors and Modeling Analysis. *Atmosphere* **2(** *11*, 456. [CrossRef]
36. Lan, X.; Thoning, K.W.; Dlugokencky, E.J. Trends in Globally-Averaged CH_4, N_2O, and SF_6 Determined from NOAA Glo Monitoring Laboratory Measurements. Version 2024-06, 2024. Available online: https://gml.noaa.gov/ccgg/trends_doi.h (accessed on 18 June 2024).
37. Laughner, J.L.; Neu, J.L.; Schimel, D.; Wennberg, P.O.; Barsanti, K.; Bowman, K.W.; Chatterjee, A.; Croes, B.E.; Fitzmaurice, H Henze, D.K.; et al. Societal shifts due to COVID-19 reveal large-scale complexities and feedbacks between atmospheric chemis and climate change. *Proc. Natl. Acad. Sci. USA* **2021**, *118*, e2109481118. [CrossRef] [PubMed]
38. Donateo, A.; Lo Feudo, T.; Marinoni, A.; Calidonna, C.R.; Contini, D.; Bonasoni, P. Long-term observations of aerosol opt properties at three GAW regional sites in the Central Mediterranean. *Atmos. Res.* **2020**, *241*, 104976. [CrossRef]
39. Grossi, G.; Goglio, P.; Vitali, A.; Williams, A.G. Livestock and climate change: Impact of livestock on climate and mitigat strategies. *Anim. Front.* **2019**, *9*, 69–76. [CrossRef]
40. Tapio, I.; Snelling, T.J.; Strozzi, F.; Wallace, R.J. The ruminal microbiome associated with methane emissions from rumin livestock. *J. Anim. Sci. Biotechnol.* **2017**, *8*, 7. [CrossRef]
41. Hristov, A.N.; Oh, J.; Firkins, J.L.; Dijkstra, J.; Kebreab, E.; Waghorn, G.; Makkar, H.P.S.; Adesogan, A.T.; Yang, W.; Lee, C.; e Special topics—Mitigation of methane and nitrous oxide emissions from animal operations: I. A review of enteric meth mitigation options. *J. Anim. Sci.* **2013**, *91*, 5045–5069. [CrossRef] [PubMed]
42. Carslaw, D.C.; Beevers, S.D.; Ropkins, K.; Bell, M.C. Detecting and quantifying aircraft and other on-airport contributions ambient nitrogen oxides in the vicinity of a large international airport. *Atmos. Environ.* **2006**, *40*, 5424–5434. [CrossRef]
43. Carslaw, D.C.; Beevers, S.D. Characterising and understanding emission sources using bivariate polar plots and k-me clustering. *Environ. Model. Softw.* **2013**, *40*, 325–329. [CrossRef]
44. Lee, D.S.; Fahey, D.; Forster, P.M.; Newton, P.J.; Wit, R.C.N.; Lim, L.L.; Owen, B.; Sausen, R. Aviation and global climate chang the 21st century. *Atmos. Environ.* **2009**, *43*, 3520–3537. [CrossRef]
45. Lee, D.S.; Fahey, D.W.; Skowron, A.; Allen, M.R.; Burkhardt, U.; Chen, Q.; Doherty, S.J.; Freeman, S.; Forster, P.M.; Fuglestvedt et al. The contribution of global aviation to anthropogenic climate forcing for 2000 to 2018. *Atmos. Environ.* **2021**, *244*, 117 [CrossRef] [PubMed]

Grobler, C.; Wolfe, P.J.; Dasadhikari, K.; Dedoussi, I.C.; Allroggen, F.; Speth, R.L.; Eastham, S.D.; Agarwal, A.; Staples, M.D.; Sabnis, J.; et al. Marginal climate and air quality costs of aviation emissions. *Environ. Res. Lett.* **2019**, *14*, 114031. [CrossRef]
Quadros, F.D.A.; Snellen, M.; Dedoussi, I.C. Regional sensitivities of air quality and human health impacts to aviation emissions. *Environ. Res. Lett.* **2020**, *15*, 105013. [CrossRef]
Barrett, S.R.H.; Britter, R.E.; Waitz, I.A. Global Mortality Attributable to Aircraft Cruise Emissions. *Environ. Sci.* **2010**, *44*, 7736–7742. [CrossRef] [PubMed]
Yim, S.H.L.; Lee, G.L.; Lee, I.H.; Allroggen, F.; Ashok, A.; Caiazzo, F.; Eastham, S.D.; Malina, R.; Barrett, S.R.H. Global, regional and local health impacts of civil aviation emissions. *Environ. Res. Lett.* **2015**, *10*, 034001. [CrossRef]
Wild, O.; Prather, M.J.; Akimoto, H. Indirect long-term global radiative cooling from NO_x emissions. *Geophys. Res. Lett.* **2001**, *28*, 1719–1722. [CrossRef]
Stevenson, D.S.; Doherty, R.M.; Sanderson, M.G.; Collins, W.J.; Johnson, C.E.; Derwent, R.G. Radiative forcing from aircraft NO_x emissions: Mechanisms and seasonal dependence. *J. Geophys. Res. Atmos.* **2004**, *109*, D17307. [CrossRef]
McNorton, J.; Bousserez, N.; Agustí-Panareda, A.; Balsamo, G.; Cantarello, L.; Engelen, R.; Huijnen, V.; Inness, A.; Kipling, Z.; Parrington, M.; et al. Quantification of methane emissions from hotspots and during COVID-19 using a global atmospheric inversion. *Atmos. Chem. Phys.* **2022**, *22*, 5961–5981. [CrossRef]
Feng, L.; Palmer, P.I.; Parker, R.J.; Lunt, M.F.; Bösch, H. Methane emissions are predominantly responsible for record-breaking atmospheric methane growth rates in 2020 and 2021. *Atmos. Chem. Phys.* **2023**, *23*, 4863–4880. [CrossRef]
Turner, A.J.; Fung, I.; Naik, V.; Horowitz, L.W.; Cohen, R. Modulation of hydroxyl variability by ENSO in the absence of external forcing. *Proc. Natl. Acad. Sci. USA* **2018**, *115*, 8931–8936. [CrossRef]
Zhao, Y.; Saunois, M.; Bousquet, P.; Lin, X.; Berchet, A.; Hegglin, M.I.; Canadell, J.G.; Jackson, R.B.; Deushi, M.; Jöckel, P.; et al. On the role of trend and variability in the hydroxyl radical (OH) in the global methane budget. *Atmos. Chem. Phys.* **2020**, *20*, 13011–13022. [CrossRef]
Nicely, J.M.; Canty, T.P.; Manyin, M.; Oman, L.D.; Salawitch, R.J.; Steenrod, S.D.; Strahan, S.E.; Strode, S.A. Changes in Global Tropospheric OH Expected as a Result of Climate Change Over the Last Several Decades. *J. Geophys. Res. Atmos.* **2018**, *123*, 10774–10795. [CrossRef]
Bândă, N.; Krol, M.; van Noije, T.; van Weele, M.; Williams, J.E.; Le Sager, P.; Niemeier, U.; Thomason, L.; Röckmann, T. The effect of stratospheric sulfur from Mount Pinatubo on tropospheric oxidizing capacity and methane. *J. Geophys. Res. Atmos.* **2015**, *120*, 1202–1220. [CrossRef]
Hossaini, R.; Chipperfield, M.P.; Saiz-Lopez, A.; Fernandez, R.; Monks, S.; Feng, W.; Brauer, P.; von Glasow, R. A global model of tropospheric chlorine chemistry: Organic versus inorganic sources and impact on methane oxidation. *J. Geophys. Res. Atmos.* **2016**, *121*, 14271–14297. [CrossRef]
Gromov, S.; Brenninkmeijer, C.A.M.; Jöckel, P. A very limited role of tropospheric chlorine as a sink of the greenhouse gas methane. *Atmos. Chem. Phys.* **2018**, *18*, 9831–9843. [CrossRef]
Saueressig, G.; Bergamaschi, P.; Crowley, J.N.; Fischer, H.; Harris, G.W. Carbon kinetic isotope effect in the reaction of CH_4 with Cl atoms. *J. Geophys. Res. Atmos.* **1995**, *22*, 1225–1228. [CrossRef]
Lan, X.; Nisbet, E.G.; Dlugokencky, E.J.; Michel, S.E. Improved Constraints on Global Methane Emissions and Sinks Using $\delta^{13}C$-CH_4. *J. Geophys. Res. Atmos.* **2021**, *35*, e2021GB007000. [CrossRef]
Le Mer, J.; Roger, P. Production, oxidation, emission and consumption of methane by soils: A review. *Eur. J. Soil Biol.* **2001**, *37*, 25–50. [CrossRef]
Hanson, R.S.; Hanson, T.E. Methanotrophic bacteria. *Microbiol. Rev.* **1996**, *60*, 439–471. [CrossRef]
Oh, Y.; Zhuang, Q.; Liu, L.; Welp, L.R.; Lau, M.C.Y.; Onstott, T.C.; Medvigy, D.; Bruhwiler, L.; Dlugokencky, E.J.; Hugelius, G.; et al. Reduced net methane emissions due to microbial methane oxidation in a warmer Arctic. *Nat. Clim. Chang.* **2020**, *10*, 317–321. [CrossRef]
Goldman, M.B.; Groffman, P.M.; Pouyat, R.V.; McDonnell, M.J.; Pickett, S.T.A. CH_4 uptake and N availability in forest soils along an urban to rural gradient. *Soil Biol. Biochem.* **1995**, *27*, 281–286. [CrossRef]
Ni, X.; Groffman, P.M. Declines in methane uptake in forest soils. *Proc. Natl. Acad. Sci. USA* **2018**, *115*, 8587–8590. [CrossRef] [PubMed]

remote sensing

MDPI

Article

Research on Multiscale Atmospheric Chaos Based on Infrared Remote-Sensing and Reanalysis Data

Zhong Wang [1,2,3], Shengli Sun [1,2,*], Wenjun Xu [1,2], Rui Chen [1,2,3], Yijun Ma [1,2,3] and Gaorui Liu [1,2]

1 Shanghai Institute of Technical Physics, Chinese Academy of Sciences, Shanghai 200083, China; wangzhong19@mails.uacs.ac.cn (Z.W.); xuwenjun@mail.sitp.ac.cn (W.X.); chenrui@mail.sitp.ac.cn (R.C.); mayijun@mail.sitp.ac.cn (Y.M.); liugaorui@mail.sitp.ac.cn (G.L.)
2 University of Chinese Academy of Sciences, Beijing 100049, China
3 Key Laboratory of Intelligent Infrared Sensing, Chinese Academy of Sciences, Shanghai 200083, China
* Correspondence: palm_sun@mail.sitp.ac.cn

Abstract: The atmosphere is a complex nonlinear system, with the information of its temperature, water vapor, pressure, and cloud being crucial aspects of remote-sensing data analysis. There exist intricate interactions among these internal components, such as convection, radiation, and humidity exchange. Atmospheric phenomena span multiple spatial and temporal scales, from small-scale thunderstorms to large-scale events like El Niño. The dynamic interactions across different scales, along with external disturbances to the atmospheric system, such as variations in solar radiation and Earth surface conditions, contribute to the chaotic nature of the atmosphere, making long-term predictions challenging. Grasping the intrinsic chaotic dynamics is essential for advancing atmospheric analysis, which holds profound implications for enhancing meteorological forecasts, mitigating disaster risks, and safeguarding ecological systems. To validate the chaotic nature of the atmosphere, this paper reviewed the definitions and main features of chaotic systems, elucidated the method of phase space reconstruction centered on Takens' theorem, and categorized the qualitative and quantitative methods for determining the chaotic nature of time series data. Among quantitative methods, the Wolf method is used to calculate the Largest Lyapunov Exponents, while the G–P method is used to calculate the correlation dimensions. A new method named Improved Saturated Correlation Dimension method was proposed to address the subjectivity and noise sensitivity inherent in the traditional G–P method. Subsequently, the Largest Lyapunov Exponents and saturated correlation dimensions were utilized to conduct a quantitative analysis of FY-4A and Himawari-8 remote-sensing infrared observation data, and ERA5 reanalysis data. For both short-term remote-sensing data and long-term reanalysis data, the results showed that more than 99.91% of the regional points have corresponding sequences with positive Largest Lyapunov exponents and all the regional points have correlation dimensions that tended to saturate at values greater than 1 with increasing embedding dimensions, thereby proving that the atmospheric system exhibits chaotic properties on both short and long temporal scales, with extreme sensitivity to initial conditions. This conclusion provided a theoretical foundation for the short-term prediction of atmospheric infrared radiation field variables and the detection of weak, time-sensitive signals in complex atmospheric environments.

Keywords: chaotic nature; atmospheric system; largest Lyapunov exponent; saturated correlation dimension; infrared remote sensing; different scales

1. Introduction

Chaos theory is a critical area within the realm of nonlinear science, focusing on the complex, irregular, and unpredictable behavior observed in deterministic systems. The origins of chaos theory can be traced back to the late 19th century. In 1892, Russian mathematician Aleksandr Lyapunov introduced the concept of the Lyapunov exponent as part of his work on analyzing the stability of dynamical systems [1]. This measure provided a way to quantify the exponential rates at which small perturbations in initial conditions either diverge or converge, offering a crucial tool for determining system stability. However, without the computational tools to numerically solve the complex equations governing such systems, the application of Lyapunov exponents was severely limited before the mid-20th century. In the early 20th century, French mathematician Henri Poincaré, while studying the three-body problem, first identified the non-periodicity of orbits and the sensitivity of systems to initial conditions [2]. Poincaré's work laid the foundation for chaos theory, revealing that even simple deterministic systems can exhibit highly complex behavior. In the 1930s, the behavior observed by Balthasar van der Pol in his studies of nonlinear circuit oscillators demonstrated the complex dynamics of systems and further revealed chaotic characteristics [3].

The significant breakthrough in chaos theory occurred in the 1960s. Meteorologist Edward Lorenz, while working on weather prediction models, accidentally discovered the system later known as the Lorenz attractor [4]. This discovery demonstrated that even simple weather models could exhibit unpredictable long-term behavior, showing extreme sensitivity to initial conditions—a phenomenon later termed the "butterfly effect". In 1967, Stephen Smale introduced the horseshoe map theory [5], providing another classic model of chaotic systems and further deepening the understanding of chaotic behavior. Furthermore, with the development of computational techniques, the ability to characterize the sensitivity to initial conditions of Lyapunov exponent has since made it a fundamental instrument in the study of chaos and complex systems.

In the 1970s, research in chaos theory entered a phase of rapid development. In 1975, James Yorke and his student introduced the concept of "Period Three Implies Chaos" [6], further enriching the mathematical foundation of chaos theory. In the same year, Mitchell Feigenbaum discovered the universal constants present in the period-doubling bifurcation process [7], unveiling the patterns through which nonlinear systems transition to chaos. This is a discovery with broad applicability. In 1976, Robert May published research on population models [8], revealing that complex chaotic behavior could also emerge in simple biological models.

In the 1980s, chaos theory experienced significant development and application. Floris Takens introduced the Embedding Theorem [9], which provided a method to reconstruct the phase space of a dynamical system using time series data from a single observable variable. Benoit Mandelbrot's fractal geometry [10] provided a new perspective for studying chaotic systems, illustrating the nature of chaotic behavior through complex fractal structures. Peter Grassberger and Itamar Procaccia introduced the concept of correlation dimension and developed the Grassberger–Procaccia (G–P) algorithm to compute this dimension from time series data [11], making it possible to characterize the complexity and quantify the fractal structure of chaotic systems. Simultaneously, the concept of strange attractors gained widespread use, aiding scientists in better understanding the long-term behavior of complex systems. Chaos theory began to be widely applied across various scientific and engineering fields, including biology, economics, physics, circuit design and meteorology.

Atmospheric systems exhibit similar chaotic behavior to the Lorenz system due to the interactions among internal components, external disturbances from solar and terrestrial radiation, and the complex influences of various variables across different temporal and spatial scales. These systems, driven by complete physical mechanisms, present pseudo-random phenomena due to their intricate structure and dynamics.

The study of the chaotic properties of the atmosphere, as a core component of the Earth's climate system, is of great significance for better understanding global climate change and predicting meteorological disasters. In recent years, the rapid advancement of satellite remote-sensing technology has enabled us to obtain abundant and precise meteorological data from the atmosphere, greatly enhancing our understanding and predictive capabilities regarding atmospheric dynamics and climate change. Radiometric imagers, by detecting radiation at different wavelengths, provide detailed information on temperature, water vapor, and cloud distribution [12]. These data play a vital role in weather forecasting, climate monitoring, and environmental protection and offer valuable resources for atmospheric science research.

Satellite remote-sensing data are categorized into different levels based on the degree of processing of the raw data. In engineering practice, Level 1 (L1) satellite data is a commonly used foundational data type. It refers to raw observation data that have undergone preliminary processing, including radiometric calibration, geometric correction, and georeferencing. L1 data are significant in studying chaotic properties of the atmosphere, as they reflect the true state and complexity of the atmospheric system. The high temporal resolution and accuracy of L1 data enable the capture of minor disturbances and changes in the atmospheric system, providing rich data sources for chaotic analysis.

Besides L1 satellite data, reanalysis data is another critical tool for atmospheric research. Reanalysis data combines satellite remote-sensing data, ground-based observations, and numerical weather prediction models to generate high spatial resolution, long-term continuous global meteorological datasets using advanced data assimilation techniques. Reanalysis data fill observation gaps and provide homogenized meteorological data, offering indispensable support for studying long-term trends in chaotic properties of the atmosphere [13]. By integrating multi-source observational data, reanalysis data eliminate uncertainties and errors from single observation methods.

In this study, we investigate the chaotic properties of the atmosphere using L1 data from FY-4A and Himawari-8 and the ERA5 reanalysis dataset from the European Centre for Medium-Range Weather Forecasts (ECMWF). For L1 satellite data, a 2–3 months time frame is selected because this period allows for the capture of short-term chaotic behaviors such as synoptic-scale weather patterns, including the development and evolution of cyclones, anticyclones, and frontal systems. Satellite data are well-suited for short-term analysis due to their high temporal and spatial resolution, which is critical for resolving these transient atmospheric phenomena. For ERA5, a 5–10 years period is appropriate because reanalysis data is designed to provide a consistent and continuous record of the atmosphere over extended periods. It allows for the identification of chaotic behavior at larger scales, such as those related to persistent anomalies, large-scale teleconnection patterns, and climate oscillations.

Section 2 systematically reviews the definition and main characteristics of chaos, discussing phase space reconstruction methods centered on Takens' theorem. Section 3 categorizes and summarizes qualitative and quantitative methods for determining the chaotic nature of time series data. To compensate for the shortcomings of the traditional G–P algorithm, a new method named Improved Saturated Correlation Dimension method was proposed to address the subjectivity and noise sensitivity when determining the scaling region. Section 4 employs quantitative determination methods, specifically the largest Lyapunov exponent and saturated correlation dimension, to analyze the chaotic nature of FY-4A and Himawari-8 remote-sensing data and ERA5 reanalysis data. In this section, we focus on the short-term chaotic nature of the atmosphere using infrared band data from L1 satellite data, while T850 and Z500 reanalysis data from ERA5, supplementing L1 data, further validate the long-term chaotic nature of the atmosphere. These analyses demonstrate that the atmospheric system exhibits chaotic behavior across different time scales. Finally, Section 5 provides prospects for future applications, including short-term prediction of atmospheric infrared radiation fields and the detection of weak time-sensitive signals in complex atmospheric environments, based on the chaotic properties of the atmosphere.

2. Overview of Chaos Theory

2.1. Definition and Key Characteristics of Chaos

In nonlinear science, chaos refers to the unpredictable, seemingly random behavior exhibited by deterministic dynamical systems due to their extreme sensitivity to initial conditions. In a chaotic system, the evolution of the system is governed by deterministic differential equations or maps. However, even minute changes in initial conditions can lead to vastly different trajectories, rendering long-term prediction impossible and causing the system to display apparently random behavior. This randomness does not originate from external disturbances but rather from the internal nonlinear coupling of the system [14]. At the same time, chaotic systems are globally stable.

The evolution of chaotic systems represents a unification of large-scale disorder and small-scale order. Their trajectories typically exhibit fractal structures, characterized by infinite levels of self-similarity and structural invariance under scaling transformations. This fractal structure reflects the complexity within the system and the features of nonlinear coupling.

2.2. Phase Space Reconstruction

Phase space reconstruction is a method used to analyze and understand nonlinear dynamical systems, particularly when observational data are limited or it is difficult to directly measure all state variables. Phase space reconstruction techniques are crucial steps in nonlinear time series analysis and form the basis for predicting chaotic systems.

In nonlinear dynamics, phase space is an abstract mathematical space describing the states of a dynamical system, where each point represents the complete state of the system at a specific moment.

When studying complex systems, often only one or a few state variables can be measured over time. The phase space reconstruction method, based on Takens' theorem, reconstructs the chaotic system's time series in a topologically equivalent sense by using delay coordinates. This method maps the system's dynamic behavior into a high-dimensional phase space, revealing its inherent deterministic structure [15].

Takens' theorem is a fundamental theorem in dynamical systems theory that provides the mathematical foundation for reconstructing the phase space of a system from time series data. Proposed by Floris Takens in 1981, this theorem addresses how to reconstruct the phase space through delay coordinate methods when complete state information is lacking.

Let x(t) be a one-dimensional time series representing the state of a dynamical system. Takens' theorem states that by introducing a time delay τ and an embedding dimension m, a new m-dimensional state vector $y(t)$ can be constructed. This new state vector can capture the geometric structure of the original phase space:

$$\mathbf{y}(t) = [x(t), x(t-\tau), x(t-2\tau), \ldots, x(t-(m-1)\tau)] \tag{1}$$

Here, τ is the time delay, typically chosen as a multiple of the system's characteristic time scale, while m is the embedding dimension, which theoretically should satisfy $m \geq 2d+1$, where d is the dimension of the dynamical system [9].

Common methods for determining the time delay τ include the Autocorrelation Function, the Average Displacement, the Mutual Information, and the Complex Autocorrelation [16]. The key is to ensure that the time-delayed series can serve as independent coordinates. Methods for determining the embedding dimension m include the Geometric Invariants, the False Nearest Neighbors, and the improved Cao [17].

Determining appropriate embedding dimensions and time delays is a crucial issue in phase space reconstruction. By selecting suitable time delays and embedding dimensions, one can reconstruct the phase space of a dynamical system under incomplete observation conditions [18].

In this study, we calculate the time delay τ using the autocorrelation method and the embedding dimension m using the False Nearest Neighbors method.

2.2.1. Determining Time Delay τ Using the Autocorrelation Method

The autocorrelation function $R(\tau)$ is used to describe the correlation of a time series at different time lags. For a given time series $\{x(t)\}$, the autocorrelation function $R(\tau)$ is defined as follows:

$$R(\tau) = \frac{\sum_{t=1}^{N-\tau} (x(t) - \overline{x})(x(t+\tau) - \overline{x})}{\sum_{t=1}^{N} (x(t) - \overline{x})^2} \tag{2}$$

The appropriate time delay τ can be determined as the time at which the autocorrelation function $R(\tau)$ first drops below $\frac{1}{e}$. This ensures that the time-delayed variables are sufficiently decorrelated for effective phase space reconstruction.

2.2.2. Determining the Embedding Dimension m Using the False Nearest Neighbors Method

The False Nearest Neighbors (FNN) method determines the embedding dimension by detecting false nearest neighbors in high-dimensional space [19]. For a given time sequence $\{x(t)\}$, after determining the time delay τ using the Autocorrelation method, increment the embedding dimension m and perform phase space reconstruction using Formula (1). The nearest neighbor distances are then calculated in both m-dimensional and m + 1-dimensional spaces:

$$\begin{aligned} d_m(t) &= \|\mathbf{X}(t) - \mathbf{X}_{NN}(t)\| \\ d_{m+1}(t) &= \|\mathbf{X}_{m+1}(t) - \mathbf{X}_{NN,m+1}(t)\| \end{aligned} \tag{3}$$

If the distance growth ratio exceeds a certain threshold:

$$R(t) = \frac{d_{m+1}(t)}{d_m(t)} > R_{th} \tag{4}$$

Then, the nearest neighbor in m-dimensional space is no longer considered the nearest neighbor in m + 1-dimensional space, indicating it is a false nearest neighbor. As m increases, when the proportion of false nearest neighbors drops to a sufficiently low level, the increment of the embedding dimension stops. The current dimension is then considered the appropriate embedding dimension m.

3. Methods for Determining Chaotic Nature

To determine whether a nonlinear system is chaotic, qualitative or quantitative methods can be used to analyze the intrinsic properties of its time series, such as phase diagrams, Lyapunov exponents, and correlation dimensions. Qualitative methods include the Phase Diagram method, the Power Spectrum method, and the Poincare Surface of Section method, while quantitative methods include the Lyapunov Exponent (LE) method, the Kolmogorov Entropy method, and the Saturated Correlation Dimension method [20]. This section will mainly discuss two of the quantitative methods which will be used in Section 4. One is the Lyapunov Exponent method. The other is the Improved Saturated Correlation Dimension method proposed in this article, which is based on curvature and the least squares method. This new method has addressed the subjectivity and noise sensitivity inherent in the traditional G–P method when determining the scaling region of the $\log C(r) - \log r$ plot.

3.1. The Lyapunov Exponent Method

In chaos theory, Lyapunov exponents (LE) are key tools for quantifying the sensitivity of a system to initial conditions [21]. Lyapunov exponents represent the rate at which nearby trajectories in the phase space diverge or converge. The formula for calculating LE is as follows:

$$LE = \lim_{n\to\infty} \frac{1}{n} \sum_{m=0}^{n-1} \ln\left|\frac{df(x_m, \mu)}{dx}\right| \tag{5}$$

where n is the time (number of iterations).

The number of Lyapunov exponents is generally equal to the dimension of the reconstructed phase space, so calculating Lyapunov exponents is based on phase space reconstruction. When the largest Lyapunov exponent (LLE) is positive, it indicates that nearby trajectories in the phase space will exponentially diverge over time, and the system exhibits chaotic behavior. When all LEs are negative, it indicates that nearby trajectories converge over time, and the system exhibits a large-scale periodic state. When LE is zero, it indicates that the distance between nearby trajectories remains unchanged over time, which is a critical state between chaos and periodicity [22]. Therefore, the sign of the largest Lyapunov exponent (LLE) can be used to determine the state of the system.

Methods for calculating LLE include the direct method, the Jacobian method, the Wolf method, and the Small Data method [23]. This section will mainly discuss the Wolf method.

Assume the chaotic time series is $x_1, x_2, \ldots, x_n$, with embedding dimension m and time delay τ. After reconstructing the phase space, a point on the attractor $Y(t_i) = \left(x(t_i), x(t_{i+\tau}), \ldots, x\left(t_{i+(m-1)\tau}\right)\right)(i = 1, 2, \ldots, N)$ is taken as a point on the reference trajectory. The initial point on the reference trajectory $Y(t_0)$ and its nearest neighbor point outside the reference trajectory $Y_0(t_0)$ are taken as initial points, with an initial distance L_0. As the two points evolve over time to t_1, the distance $L'_0 = ||Y(t_1) - Y_0(t_1)||$ exceeds a certain threshold $\varepsilon > 0$. At this point, a new nearest neighbor point $Y_1(t_1)$ is selected outside the reference trajectory, and the angle between the vectors of the new data point and the old data point with the corresponding point on the reference trajectory should be as small as possible [24]. This process is repeated until the reference trajectory traverses all the data in the time series, with a total number of iterations M. The LLE is then calculated as follows:

$$LLE = \frac{1}{t_M - t_0} \sum_{i=0}^{M} \ln \frac{L'_i}{L_i} \tag{6}$$

3.2. The Improved Saturated Correlation Dimension Method

In 1983, Grassberger and Procaccia proposed the G–P algorithm for directly calculating the correlation dimension from time series using embedding theory and phase space reconstruction techniques [25].

The correlation dimension is a measure of the degree of interdependence among variables in a complex system or dataset. It can be used to describe the complexity of attractors in the system's phase space and to identify chaos. The correlation dimension is also a type of fractal dimension, characterized by conservativeness, simplicity of calculation, and stability [26]. The process for calculating the saturated correlation dimension is as follows:

First, perform phase space reconstruction for the time series.

Second, calculate the correlation integral. The correlation integral $C(r)$ is a statistical measure of the distances between pairs of points in the phase space. For each pair of points $\mathbf{X}_i$ and $\mathbf{X}_j$ (where $i \neq j$), calculate the distance between them:

$$d_{ij} = ||\mathbf{X}_i - \mathbf{X}_j|| \tag{7}$$

The correlation integral $C(r)$ is defined as the proportion of point pairs within a given distance r:

$$C(r) = \frac{2}{N(N-1)} \sum_{i<j} H(r - d_{ij}) \tag{8}$$

where $H(x)$ is the Heaviside function, and N is the number of reconstructed phase space vectors.

Third, estimate the correlation dimension. The correlation dimension D_2 is estimated as the slope of the log–log plot of the correlation integral $C(r)$ versus the distance r in the scaling region [11]

$$\log C(r) \approx D_2 \log r + \text{const} \tag{9}$$

The scaling region refers to a range of distances r over which the correlation integral $C(r)$ exhibits a power-law relationship with r. In this region, the log–log plot appears as a straight line, indicating a linear relationship between these logarithmic values.

One of the traditional methods to identify the scaling region is to visually inspect the log–log plot to identify a linear segment [27]. However, this method involves a degree of subjectivity, which can lead to variability in the results depending on the researcher's judgment.

Another method is to estimate the slope in the region where r approaches 0 [28].

$$D_2 = \lim_{r \to 0} \frac{\log C(r)}{\log r} \tag{10}$$

The approach is grounded in the idea that, at sufficiently small scales, the system's geometry should dominate, allowing for a more accurate determination of the fractal dimension. This method reduces the reliance on visual inspection for identifying the scaling region, making the estimation of the correlation dimension more objective. However, the method can be sensitive to noise, especially in the small r region, where the behavior of $\log C(r)$ may be distorted by data imperfections.

In order to address the subjectivity and noise sensitivity inherent in the aforementioned methods, this paper proposes a curvature-based approach for determining the scaling region, which can be effectively carried out through a systematic analysis of the first and second derivatives of the log–log plot.

The first derivative of $\log C(r)$ with respect to $\log r$ gives the local slope $D(r)$, which can be interpreted as the correlation dimension D_2 in the linear region:

$$D(r) = \frac{d \log C(r)}{d \log r} \tag{11}$$

This derivative can be numerically approximated using the finite difference method:

$$D(r_i) \approx \frac{\log C(r_{i+1}) - \log C(r_i)}{\log r_{i+1} - \log r_i} \tag{12}$$

The second derivative, representing the curvature of the log–log plot, is instrumental in identifying deviations from linearity:

$$\text{Curvature } (r) = \frac{d^2 \log C(r)}{d(\log r)^2} \tag{13}$$

Numerically, this can be approximated as:

$$\text{Curvature}(r_i) \approx \frac{D(r_{i+1}) - D(r_i)}{\log r_{i+1} - \log r_i} \tag{14}$$

The scaling region is identified by examining the stability of $D(r)$ and minimal curvature. The range of r values where the first derivative $D(r)$ remains relatively constant, indicating a stable slope, suggests a linear region in the log–log plot. The region where the second derivative (curvature) is close to zero further supports the presence of a linear relationship. The overlap between these regions indicates the appropriate scaling region.

By analyzing the stability of the slope (first derivative) and minimizing the curvature (second derivative), this method ensures a reliable determination of the scaling region, facilitating a more accurate estimation of the correlation dimension.

After determining the scaling region, the slope of $\log C(r)$ over $\log r$ is calculated using the least squares method to obtain the correlation dimension D_2.

Fourth, verify saturation. Calculate the correlation dimensions D_2 for different embedding dimensions m and plot the variation curve with m as the horizontal axis and D_2 as the vertical axis.

The correlation dimension is essentially a type of fractal dimension used to describe the complexity of an attractor. A fractal dimension greater than 1 indicates that the attractor possesses a complex geometric structure, which is characteristic of chaotic systems. For periodic or non-chaotic systems, their attractors are typically low-dimensional, such as fixed points, periodic orbits, or slightly more complex quasi-periodic structures. The correlation dimension of these systems is usually close to or equal to 1, as they do not exhibit fractal structures [29,30]. Therefore, if the correlation dimension saturates at dimension greater than 1, the time series has chaotic characteristics. The higher the saturated correlation dimension, the stronger the chaos.

4. Analysis of Chaotic Properties in Atmospheric Data

This section will investigate the chaotic properties of the atmosphere using L1 data from FY-4A and Himawari-8, as well as reanalysis data from ERA5. Specifically, short-term chaotic properties (2 to 3 months) will be analyzed using L1 infrared and visible band data of the satellites, while long-term chaotic properties (5 to 10 years) will be analyzed using T850 and Z500 reanalysis data from ERA5.

Two methods will be applied to each dataset to analyze its chaotic nature, including the Wolf method and the Improved Saturated Correlation Dimension method. The overall process is shown in Figure 1.

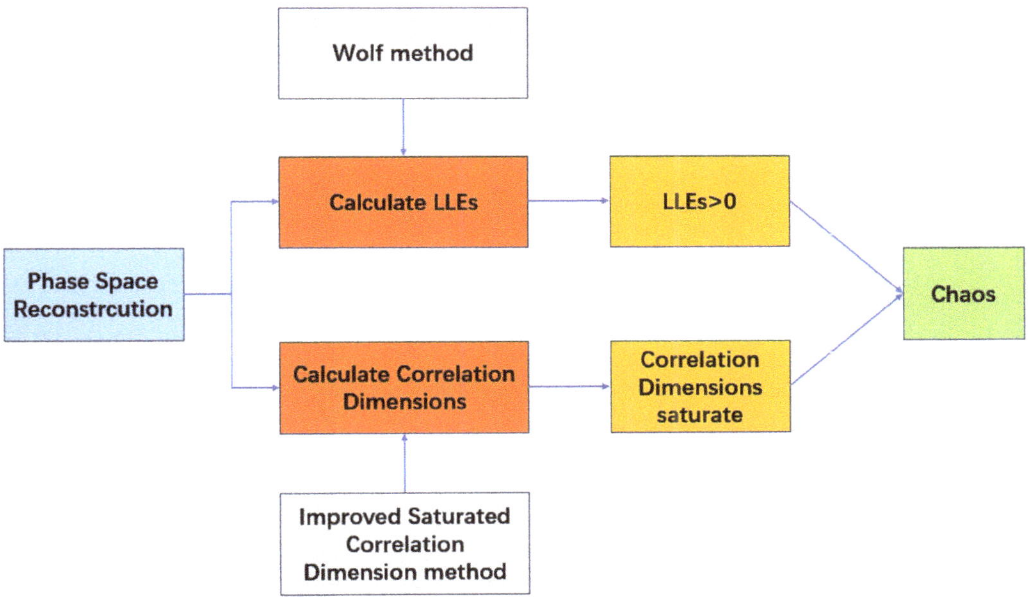

Figure 1. Analysis process of chaotic nature.

4.1. Analyzing Chaos Using Largest Lyapunov Exponents

The Wolf method for calculating the Largest Lyapunov Exponent (LLE) is a numerical technique used to estimate the separation rate of nearby trajectories in a dynamical system, and it is a method for quantitatively determining the chaotic nature of time series data.

The following sections will use phase space reconstruction and the Wolf method to calculate LLEs for the AGRI China region 4KM L1 data from FY-4A, the AHI full-disk 2KM L1 data from Himawari-8, and the Z500 and T850 data from ERA5. This will allow for a quantitative analysis of their short-term and long-term chaotic properties.

4.1.1. LLEs of FY-4A AGRI 4KM L1 Data

Using the FY-4A Advanced Geostationary Radiation Imager (AGRI) data for the China region at 4KM L1 resolution, the focus is on channel 1 (visible band, central wavelength 0.47 microns), channel 4 (shortwave infrared band, central wavelength 1.375 microns), channel 10 (midwave infrared band, central wavelength 7.1 microns), and channel 13 (longwave infrared band, central wavelength 12.0 microns). These specific channels were selected because they correspond to the visible light, shortwave infrared, midwave infrared, and longwave infrared bands. Channels 1, 4, 10, and 13 are typically used for observing color composite images, cirrus clouds, water vapor, and temperature, respectively. These factors are key drivers of atmospheric dynamics. The combination of these channels can provide a broad perspective of atmospheric processes.

Data were extracted for a 60×60 spatial coordinate area, totaling 3600 points, with a spatial resolution of 4 km. For each point, a time series of 90 days with a temporal resolution of 165 times per day was captured, resulting in a sequence with a total length of 14,850. This provided a $60 \times 60 \times 14{,}850$ three-dimensional spatiotemporal dataset.

Phase space reconstruction is performed on the time series for all coordinate points in this region. The time delay τ and embedding dimension m are calculated using the Autocorrelation method and the False Nearest Neighbors method, respectively, resulting in 60×60 matrices of time delays and embedding dimensions. It was found that all m values are equal to 8, but the τ values vary among points. Thus, phase space reconstruction is conducted for each point's time series based on its corresponding τ and m values using Takens' theorem. The Wolf method is then employed to calculate the LLEs. LLEs corresponding to each band in this region are shown in Figure 2. The x-axis and y-axis represent the spatial coordinates in the longitude and latitude directions, respectively. If the value at a certain point is greater than 0, it indicates that the LLE at that point is positive. This means that the time series at that location will diverge over time, exhibiting chaotic behavior. When equal to 0, the time series corresponds to quasi-periodic motion or a boundary state in the system, where the behavior is regular but not chaotic. A negative LLE means that the series tends to converge over time, implying that it is stable.

Using the above method, the data range was extended to cover the entire region and all bands, and the corresponding LLEs were calculated. According to the results, for the FY-4A AGRI China region 4KM L1 data across all bands, more than 99.91% (5,381,233 out of 5,386,080 points, which is the total number of regional points of the data) of the regional points have a corresponding sequence with a positive LLE. Only less than 0.09% of the regional points have a corresponding sequence with a negative LLE. These negative LLEs may result from surface influences such as infrared emission and reflection from the Earth's surface, terrain-induced atmospheric stability, and boundary layer interactions, which can create localized areas of reduced dynamical complexity and stability in the atmospheric system. On the other hand, negative LLEs may also be attributed to potential numerical artifacts arising from resolution constraints or boundary effects of satellite image. The result above leads to the conclusion that the infrared and visible radiation fields of the atmospheric medium exhibit chaotic nature in the short term.

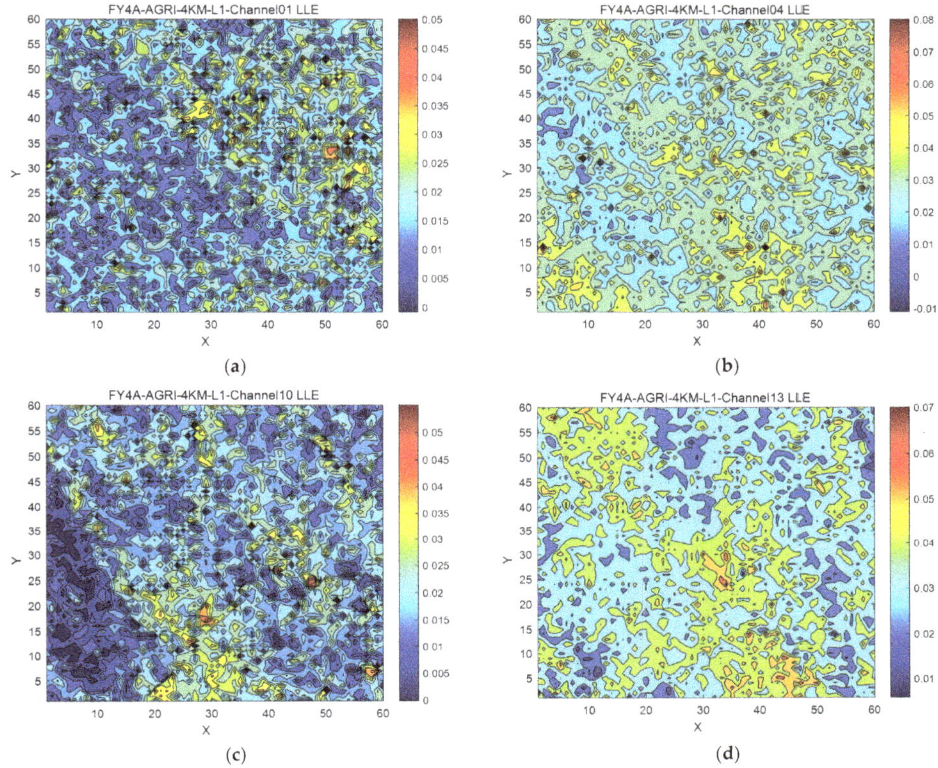

Figure 2. (**a**) 60 × 0 LLE diagram of FY-4A AGRI China Area 4KM L1 data for Channel 1; (**b**) 60 × 60 LLE diagram of FY-4A AGRI China Area 4KM L1 data for Channel 4; (**c**) 60 × 60 LLE diagram of FY-4A AGRI China Area 4KM L1 data for Channel 10; (**d**) 60 × 60 LLE diagram of FY-4A AGRI China Area 4KM L1 data for Channel 13.

4.1.2. Himawari-8 AHI 2KM L1 Data

Using the Himawari-8 Advanced Himawari Imager (AHI) full-disk 2KM L1 data for channel 2 (visible band, central wavelength 0.51 microns) and channel 12 (infrared band, central wavelength 9.6 microns) as an example, data were extracted for a 40 × 40 spatial coordinate area, totaling 1600 points [31]. For each point, a time series of 61 days with a temporal resolution of 10 min was captured, resulting in a sequence with a total length of 8784. This provided a 40 × 40 × 8784 three-dimensional spatiotemporal dataset.

Phase space reconstruction is performed on the time series for all coordinate points in this region. The time delay τ and embedding dimension m are calculated using the Autocorrelation method and the False Nearest Neighbors method. The embedding dimensions for all sequences in this region are consistent, with $m = 8$. The time delays for channel 2 range from 12 to 14, and for channel 12, they range from 30 to 38. Based on the embedding dimensions and the time delays corresponding to each sequence, phase space reconstruction is conducted for all sequences using Takens' theorem. The Wolf method is then employed to calculate LLEs. LLEs corresponding to each band in this region are shown in Figure 3. The x-axis and y-axis represent the spatial coordinates in the longitude and latitude directions, respectively.

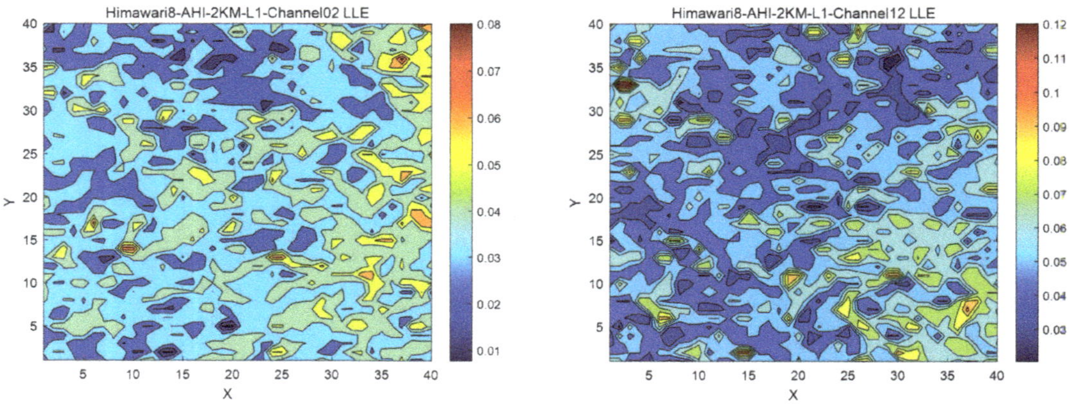

(a) (b)

Figure 3. (**a**) 40 × 40 LLE diagram of Himawari-8 AHI Full Disk 2KM L1 data for Channel 2; (**b**) 40 × 40 LLE diagram of Himawari-8 AHI Full Disk 2KM L1 data for Channel 12.

Using the above method, the data range was extended to a global scale and all bands, and the corresponding LLEs were calculated. According to the results, for the Himawari-8 satellite's AHI full-disk 2KM L1 data across all bands, 99.97% of the regional points have a corresponding sequence with a positive LLE. This leads to the conclusion that the infrared and visible radiation field of the atmospheric medium exhibit chaotic nature in the short term.

4.1.3. ERA5 Z500 and T850 Data

Using the ERA5 reanalysis dataset from the European Centre for Medium-Range Weather Forecasts (ECMWF), specifically the 500 hPa geopotential height data and the 850 hPa temperature data as examples, data were extracted with a spatial resolution of 0.25 degrees. This corresponds to a 41 × 41 grid, covering a 10-degree by 10-degree latitude–longitude area, totaling 1681 points. For each point, a time series from 2015 to 2021 with an hourly resolution was captured, resulting in a total length of 61,368 sequences. This provided a 41 × 41 × 61,368 three-dimensional spatiotemporal dataset.

Phase space reconstruction is performed on the time series for all coordinate points in this region. The time delay τ and embedding dimension m are calculated using the autocorrelation method and the false nearest neighbors method. The time delays and embedding dimensions for all sequences in this region are nearly consistent. For Z500 data, $\tau = 4$ and $m = 8$. For T850 data, $\tau = 5$ and $m = 8$. Based on these values, phase space reconstruction is conducted for all sequences using Takens' theorem, and the Wolf method is employed to calculate the LLE. LLEs corresponding to each band in this region are shown in Figure 4.

Using the above method, the data range was extended to a global scale, and the corresponding LLEs were calculated. According to the results, for ERA5 Z500 and T850 data, all the points have a corresponding sequence with a positive LLE. This leads to the conclusion that the atmospheric medium exhibits chaotic nature in the long term.

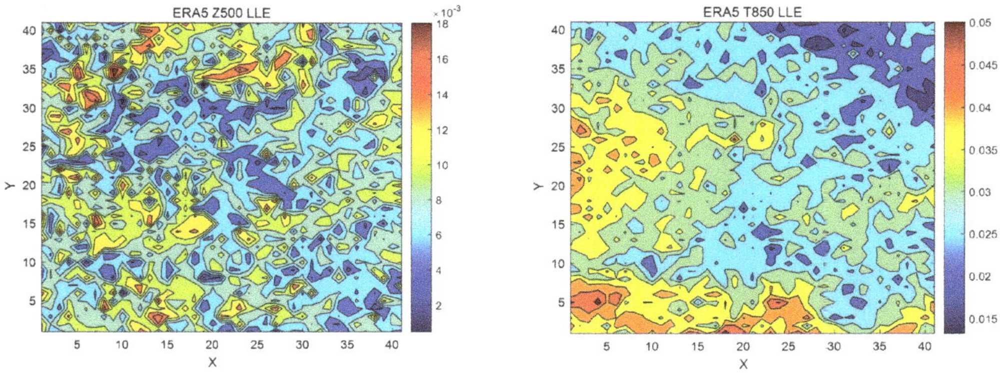

(a) (b)

Figure 4. (**a**) 41 × 41 LLE diagram of ERA5 hourly data for Z500; (**b**) 41 × 41 LLE diagram of ERA5 hourly data for T850.

4.2. Analyzing Chaos Using Correlation Dimensions

Using the Improved Saturated Correlation Dimension method described in Section 3.2, the correlation integral for the data in Section 4.1 was calculated and the $\log C(r) - \log r$ curve was plotted. Both the first derivative and the second derivative were calculated to obtain the scaling region. Linear regression was used to calculate the slope in that region, which represents the correlation dimension D_2. By calculating the correlation dimensions corresponding to different embedding dimensions m, the D_2-m curve was obtained. Whether D_2 tends to saturate as m increases and the specific value of D_2 when saturation occurs, determines whether the sequence is chaotic.

The above method was applied to the FY-4A AGRI 4KM L1, Himawari-8 AHI 2KM L1, ERA5 Z500, and T850 data to analyze the chaotic properties of the atmosphere on both long-term (5 to 10 years) and short-term (2 to 3 months) scales.

4.2.1. Correlation Dimensions of FY-4A AGRI 4KM L1 Data

For the FY-4A AGRI China region 4KM L1 data at each grid point in each channel, data were extracted for a time length of 90 days with a temporal resolution of 165 times per day, resulting in a total sequence length of 14,850. Channel 1 (visible band, central wavelength 0.47 microns), channel 4 (shortwave infrared band, central wavelength 1.375 microns), channel 10 (midwave infrared band, central wavelength 7.1 microns), and channel 13 (longwave infrared band, central wavelength 12.0 microns) were taken as examples to calculate the correlation dimensions. Utilizing the G–P method described in 3.2 (see Equations (7) and (8)), the log–log plots of the correlation integral $C(r)$ versus the distance r for different embedding dimensions m were plotted as Figure 5a–d, corresponding to these 4 channels of FY-4A, respectively. Then, the correlation dimensions were calculated using the curvature-based method proposed in Section 3.2 (see Equations (12) and (14)). The results obtained were largely consistent with the scaling region identified through subjective observation, which validated the feasibility and accuracy of this method. Furthermore, the D_2-m curves of these 4 channels were plotted as Figure 6. From Figure 6, the consistency of the estimated correlation dimension across different embedding dimensions has been verified, which further demonstrated the robustness and accuracy of this curvature-based method.

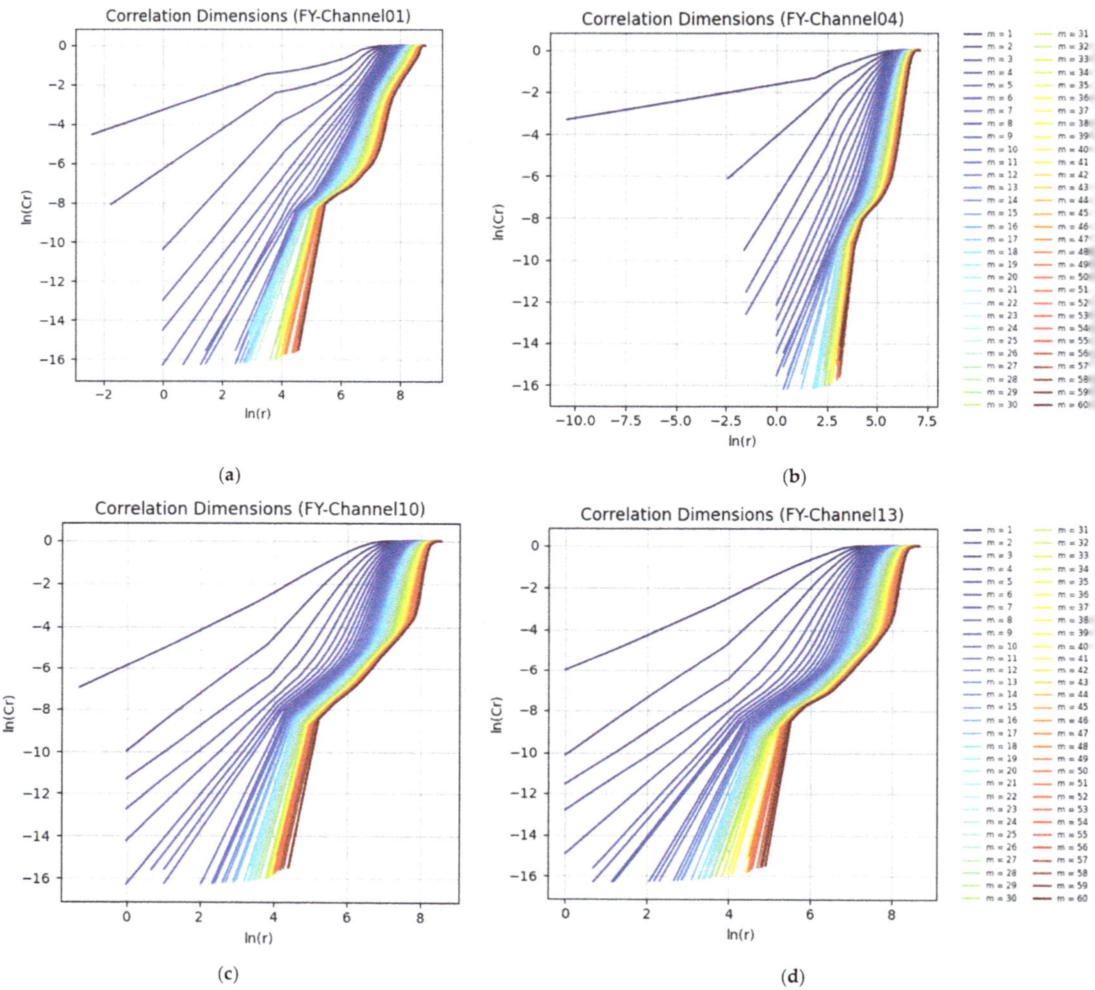

Figure 5. (**a**) Log–log plots of the correlation integral $C(r)$ versus the distance r for different embedding dimensions m for channel 1 of FY-4A AGRI China area 4KM L1 data; (**b**) Log–log plots of the correlation integral $C(r)$ versus the distance r for different embedding dimensions m for channel 4 of FY-4A AGRI China area 4KM L1 data; (**c**) Log–log plots of the correlation integral $C(r)$ versus the distance r for different embedding dimensions m for channel 10 of FY-4A AGRI China area 4KM L1 data; (**d**) Log–log plots of the correlation integral $C(r)$ versus the distance r for different embedding dimensions m for channel 13 of FY-4A AGRI China area 4KM L1 data of FY-4A AGRI China area 4KM L1 data.

Finally, the second derivatives of the D_2-m curves were calculated. When the second derivative is smaller than a small threshold, it indicates that the correlation dimension D_2 is approaching saturation. Here, the threshold was set to 0.1. From Figure 6, it can be observed that the correlation dimensions for channels 1, 4, 10, and 13 tend to saturate at approximately 3.2, 3.7, 3.2 and 3.5, respectively, as the embedding dimension increases. This saturation indicates that the sequences exhibit chaotic nature. Similarly, the correlation dimensions calculated for data from each channel and each grid point ultimately saturated at dimensions greater than 1 as the embedding dimension increases, proving that the FY-4A AGRI 4KM L1 data exhibits chaotic nature on a short-term scale.

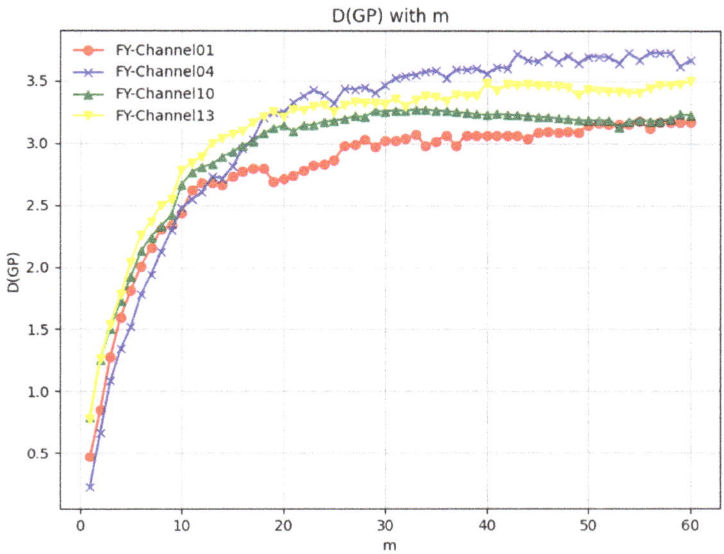

Figure 6. Correlation Dimension vs. Embedding Dimension Curve for channel 1, channel 4, channel 10 and channel 13 of FY-4A AGRI China area 4KM L1 data.

4.2.2. Correlation Dimensions of Himawari-8 AHI 2KM L1 Data

For the Himawari-8 AHI full-disk 2KM L1 data at each grid point in each channel, data were extracted for a time length of 61 days with a temporal resolution of 10 min, resulting in a total sequence length of 8784. Grid point at coordinates (4050, 5250) (approximately 20.98°S, 175.02°W) from channel 2 (visible band, central wavelength 0.51 microns) and channel 12 (infrared band, central wavelength 9.6 microns) were taken as examples to calculate the correlation dimensions. Utilizing the G–P method described in 3.2 (see Equations (7) and (8)), the log–log plots of the correlation integral $C(r)$ versus the distance r for different embedding dimensions m were plotted as Figure 7a,b, corresponding to these 2 channels of Himawari-8, respectively. Then, the correlation dimensions were calculated using the curvature-based method proposed in 3.2 (see Equations (12) and (14)). The results obtained were largely consistent with the scaling region identified through subjective observation. The D_2-m curves were plotted as Figure 8. The consistency of the estimated correlation dimension across different embedding dimensions has been verified in Figure 8, which further demonstrated the feasibility and accuracy of this curvature-based method.

From Figure 8, it can be observed that the correlation dimensions for channel 2 and 12 tended to saturate at approximately 5.1 and 4.9, respectively, where the second derivatives of the D_2-m curves were smaller than the threshold 0.1. This saturation indicates that the sequences exhibit chaotic nature. Similarly, the correlation dimensions calculated for data from each channel and each grid point ultimately saturated at dimensions greater than 1 as the embedding dimension increases, proving that the Himawari-8 AHI full-disk 2KM L1 data exhibits chaotic nature on a short-term scale.

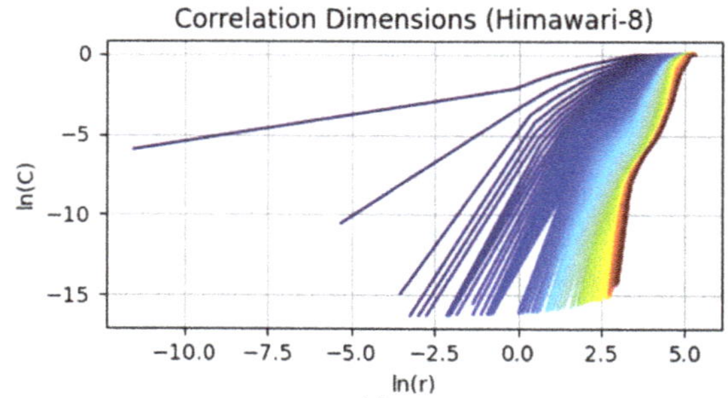

(**a**)

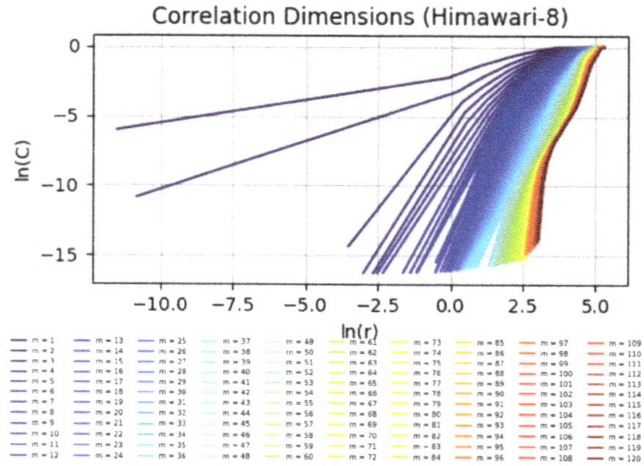

(**b**)

Figure 7. (**a**) Log–log plots of the correlation integral $C(r)$ versus the distance r for different embedding dimensions m for channel 2 of Himawari-8 AHI full-disk 2KM L1 data; (**b**) Log–log plots of the correlation integral $C(r)$ versus the distance r for different embedding dimensions m for channel 12 of Himawari-8 AHI full-disk 2KM L1 data.

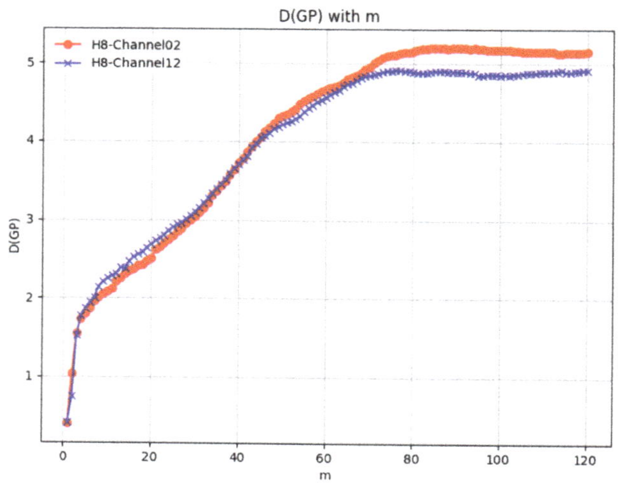

Figure 8. Correlation Dimension vs. Embedding Dimension Curve for channel 2 and channel 12 of Himawari-8 AHI Full disk 2KM L1 data at location (4050, 5250).

4.2.3. Correlation Dimensions of ERA5 Z500 and T850 Data

For the ERA5 reanalysis dataset from European Centre for Medium-Range Weather Forecasts (ECMWF), specifically the 500 hPa geopotential height data and the 850 hPa temperature data, data were extracted for the period from 2015 to 2021 with an hourly resolution, resulting in a total sequence length of 61,368. Utilizing the G–P method, described in Section 3.2 (see Equations (7) and (8)), the log–log plots of the correlation integral $C(r)$ versus the distance r for different embedding dimensions m were plotted as Figure 9a,b, corresponding to Z500 and T850 hourly data of ERA5 at 20.00°S, 175.00°W, respectively. Then, the correlation dimensions were calculated using the curvature-based method proposed in Section 3.2 (see Equations (12) and (14)). The D_2-m curves were plotted as Figure 10. The scaling regions obtained were largely consistent with those identified through subjective observation. And the consistency of the estimated correlation dimension across different embedding dimensions has been verified in Figure 10, which demonstrated the feasibility and accuracy of this curvature-based method.

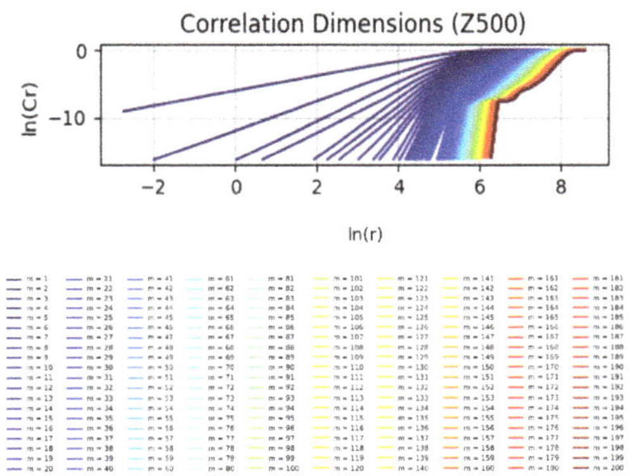

(a)

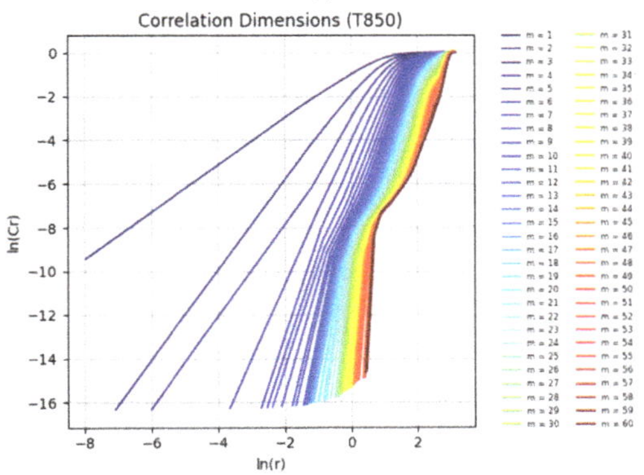

(b)

Figure 9. (**a**) Log–log plots of the correlation integral $C(r)$ versus the distance r for different embedding dimensions m for Z500 hourly data of ERA5 at 20.00°S, 175.00°W; (**b**) Log–log plots of the correlation integral $C(r)$ versus the distance r for different embedding dimensions m for T850 hourly data of ERA5 at 20.00°S, 175.00°W.

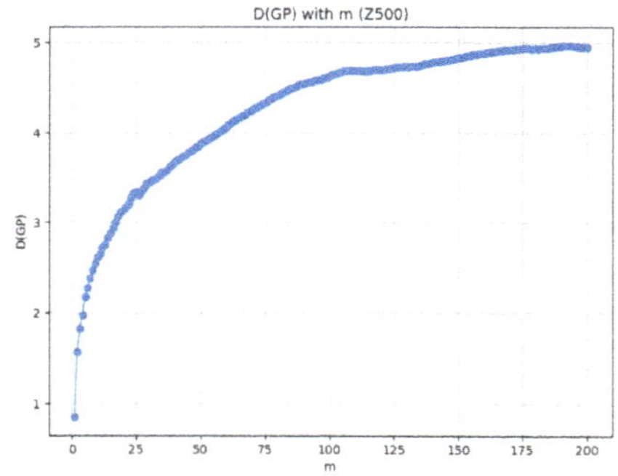

(**a**)

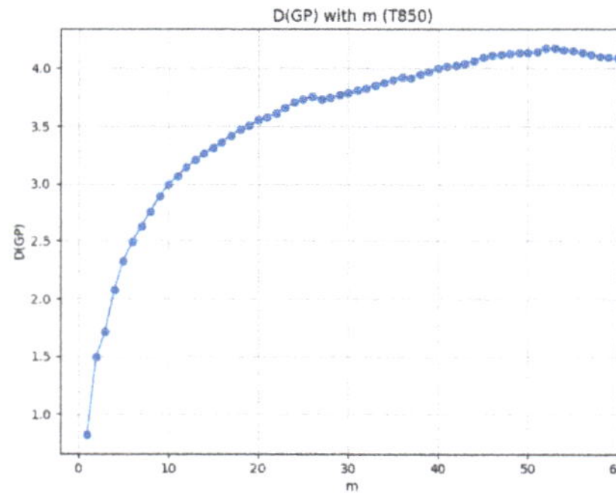

(**b**)

Figure 10. (**a**) Correlation Dimension vs. Embedding Dimension Curve for Z500 hourly data of ERA5 at 20.00°S, 175.00°W; (**b**) Correlation Dimension vs. Embedding Dimension Curve for T850 hourly data of ERA5 at 20.00°S, 175.00°W.

The correlation dimensions for the Z500 and T850 sequences tended to saturate at approximately 4.95 and 4.05, respectively, as the embedding dimension increases, demonstrating the chaotic nature of these sequences. Similarly, the correlation dimensions calculated for sequences at each latitude and longitude point ultimately saturate at dimensions greater than 1 as the embedding dimension increases, proving that the ERA5 Z500 and T850 hourly data exhibit chaotic nature on a long-term scale.

5. Conclusions

In this paper, an overview of the definition and characteristics of chaos was provided. Phase space reconstruction, as a crucial preprocessing method for chaotic data, was introduced. And two quantitative methods for determining chaotic nature were thoroughly discussed. The first was the Lyapunov Exponent method, and the second was the Improved Saturated Correlation Dimension method, which was newly proposed in this work. This new method, based on curvature and the least squares approach, effectively addressed the subjectivity and noise sensitivity issues in determining the scaling region of the $\log C(r) - \log r$ curves associated with the traditional G–P method. On this basis, this paper innovatively applies the above two methods to satellite remote-sensing L1 datasets including FY-4A and Himawari-8, as well as ERA5 reanalysis dataset. Two indicators were analyzed: whether the system's Largest Lyapunov Exponent is positive and whether the correlation dimension tends to saturate with the embedding dimension. The results showed that, on the one hand, the overwhelming majority of sequences of satellite remote-sensing L1 dataset (99.91% for FY-4A and 99.97% for Himawari-8) and all the sequences of ERA5 have positive LLEs. Considering the surface influences from the Earth's surface, and boundary layer interactions, which may result in the negative LLEs of less than 0.01%, the results obtained from the more comprehensive ERA5 dataset can be considered reliable. On the other hand, all the sequences of both satellite remote-sensing L1 dataset and ERA5 dataset have correlation dimensions saturating at values greater than 1. Taking into account the above two indicators comprehensively, the following conclusion can be drawn: on both short-term and long-term scales, the atmospheric medium exhibits chaotic nature.

In summary, this paper demonstrated the chaotic behavior of the atmospheric system on different scales, which helps to reveal the multi-scale interactions and coupling mechanisms of the atmospheric system. This provides an experimental basis for a deeper understanding of the highly complex internal structure and dynamic evolution mechanisms of the atmospheric system. On this basis, the characteristics of chaotic systems can be utilized in the future to study the evolution of variables in the atmospheric system using numerical integration and data-driven methods, further improving the accuracy of weather forecasts. Additionally, early warning and prediction of natural disasters such as volcanic eruptions and typhoons can be conducted based on minute abnormal disturbances in the atmospheric medium. On the other hand, the sensitivity to initial conditions in chaos can be used to capture weak moving targets under the atmospheric infrared background radiation field, which is of great significance for improving the detection rate of hostile targets in military confrontations.

Author Contributions: Conceptualization, Z.W. and W.X.; methodology, Z.W. and S.S.; software, Z.W.; validation, Z.W.; formal analysis, Z.W. and W.X.; investigation, Z.W., W.X., R.C. and Y.M.; resources, Z.W. and G.L.; data curation, Z.W. and W.X.; writing—original draft preparation, Z.W.; writing—review and editing, Z.W.; visualization, Z.W.; supervision, S.S. and W.X.; project administration, S.S. All authors have read and agreed to the published version of the manuscript.

Funding: This research received no external funding.

Data Availability Statement: The original data presented in the study are openly available in ERA5 at https://cds.climate.copernicus.eu, FY-4A at https://satellite.nsmc.org.cn/portalsite/default.aspx and Himawari-8 at https://www.eorc.jaxa.jp/ptree/index.html (accessed on 8 September 2024).

Acknowledgments: The authors would like to thank the members of Key Laboratory of Intelligent Infrared Sensing, Chinese Academy of Sciences, for their substantial support on hardware.

Conflicts of Interest: The authors declare no conflict of interest.

References

Lyapunov, A.M. The general problem of the stability of motion. *Int. J. Control.* **1992**, *55*, 531–534. [CrossRef]

Poincaré, H. *New Methods of Celestial Mechanics*; National Aeronautics and Space Administration: Washington, DC, USA, 1967.

Van der Pol, B.; Van Der Mark, J. Frequency demultiplication. *Nature* **1927**, *120*, 363–364. [CrossRef]

Lorenz, E.N. Deterministic nonperiodic flow. *J. Atmos. Sci.* **1963**, *20*, 130–141. [CrossRef]

Smale, S. Differentiable dynamical systems. *Bull. Am. Math. Soc.* **1967**, *73*, 747–817. [CrossRef]

Li, T.Y.; Yorke, J.A. Period Three Implies Chaos. *Am. Math. Mon.* **1975**, *82*, 985. [CrossRef]

Feigenbaum, M.J. Quantitative universality for a class of nonlinear transformations. *J. Stat. Phys.* **1978**, *19*, 25–52. [CrossRef]

May, R.M. Simple mathematical models with very complicated dynamics. *Nature* **1976**, *261*, 459–467. [CrossRef]

Takens, F. Detecting strange attractors in turbulence. In *Dynamical Systems and Turbulence, Warwick 1980: Proceedings of a Symposium Held at the University of Warwick, 1979/80*; Springer: Berlin/Heidelberg, Germany, 2006.

Mandelbrot, B.B. *The Fractal Geometry of Nature*; WH Freeman: New York, NY, USA, 1982; Volume 1.

Grassberger, P.; Procaccia, I. Measuring the strangeness of strange attractors. *Phys. D Nonlinear Phenom.* **1983**, *9*, 189–208. [CrossRef]

Shaw, J.A.; Nugent, P.W. Physics principles in radiometric infrared imaging of clouds in the atmosphere. *Eur. J. Phys.* **2013**, *34*, S111. [CrossRef]

Hersbach, H.; Bell, B.; Berrisford, P.; Hirahara, S.; Horányi, A.; Muñoz-Sabater, J.; Nicolas, J.; Peubey, C.; Radu, R.; Schepers, D.; et al. The ERA5 global reanalysis. *Q. J. R. Meteorol.Soc.* **2020**, *146*, 1999–2049. [CrossRef]

Tsonis, A.A. *Chaos: From Theory to Applications*; Springer Science & Business Media: Berlin/Heidelberg, Germany, 2012.

Kennel, M.B.; Brown, R.; Abarbanel, H.D. Determining embedding dimension for phase-space reconstruction using a geometrical construction. *Phys. Rev. A* **1992**, *45*, 3403. [CrossRef] [PubMed]

Buzug, T.; Pfister, G. Comparison of algorithms calculating optimal embedding parameters for delay time coordinates. *Phys. D Nonlinear Phenom.* **1992**, *58*, 127–137. [CrossRef]

Rhodes, C.; Morari, M. The false nearest neighbors algorithm: An overview. *Comput. Chem. Eng.* **1997**, *21*, S1149–S1154. [CrossRef]

Kim, H.S.; Eykholt, R.; Salas, J. Nonlinear dynamics, delay times, and embedding windows. *Phys. D Nonlinear Phenom.* **1999**, *127*, 48–60. [CrossRef]

Abarbanel, H.D.; Kennel, M.B. Local false nearest neighbors and dynamical dimensions from observed chaotic data. *Phys. Rev. E* **1993**, *47*, 3057. [CrossRef]

Fraser, A.M. Information and entropy in strange attractors. *IEEE Trans. Inf. Theory* **1989**, *35*, 245–262. [CrossRef]

Ott, E. *Chaos in Dynamical Systems*; Cambridge University Press: Cambridge, UK, 2002.

Wolf, A.; Swift, J.B.; Swinney, H.L.; Vastano, J.A. Determining Lyapunov exponents from a time series. *Phys. D Nonlinear Phenom.* **1985**, *16*, 285–317. [CrossRef]

Sandri, M. Numerical calculation of Lyapunov exponents. *Math. J.* **1996**, *6*, 78–84.

Benettin, G.; Galgani, L.; Giorgilli, A.; Strelcyn, J.-M. Lyapunov characteristic exponents for smooth dynamical systems and for Hamiltonian systems; a method for computing all of them. Part 1: Theory. *Meccanica* **1980**, *15*, 9–20. [CrossRef]

Grassberger, P.; Procaccia, I. Characterization of strange attractors. *Phys. Rev. Lett.* **1983**, *50*, 346. [CrossRef]

26. Simpelaere, D. Correlation dimension. *J. Stat. Phys.* **1998**, *90*, 491–509. [CrossRef]
27. Bradley, E.; Kantz, H. Nonlinear time-series analysis revisited. *Chaos Interdiscip. J. Nonlinear Sci.* **2015**, *25*, 097610. [CrossRef] [PubMed]
28. Theiler, J. Estimating fractal dimension. *JOSA A* **1990**, *7*, 1055–1073. [CrossRef]
29. Kantz, H.; Schreiber, T. *Nonlinear Time Series Analysis*; Cambridge University Press: Cambridge, UK, 2004; Volume 7.
30. Eckmann, J.-P.; Ruelle, D. Ergodic theory of chaos and strange attractors. *Rev. Mod. Phys.* **1985**, *57*, 617. [CrossRef]
31. Bessho, K.; Hayashi, M.; Ikeda, A.; Imai, T.; Inoue, H.; Kumagai, Y.; Miyakawa, T.; Murata, H.; Ohno, T.; Okuyama, A. introduction to Himawari-8/9—Japan's new-generation geostationary meteorological satellites. *J. Meteorol. Soc. Japan. Se* **2016**, *94*, 151–183. [CrossRef]

atmosphere

MDPI

icle

patial and Temporal Variations in Rainwater Chemistry in a apid Urbanization Area of Shenzhen, China

ong Huang [1,2,*], Jingwei Yang [1,2], Chang Liu [1,2], Ruiying Jing [1,2] and Qiaohui Lu [1,2]

1 Guangdong Provincial Engineering and Technology Research Center for Water Affairs Big Data and Water Ecology, Shenzhen 518001, China
2 Shenzhen Water Planning & Design Institute Co., Ltd., Shenzhen 518001, China
* Correspondence: huangyl@swpdi.com; Tel.: +86-755-2556-3295; Fax: +86-755-2589-0439

Abstract: Studying chemical constituents in rainwater can provide insights into the origin, characteristics, concentration, and processes involved in clearing atmospheric pollutants. This study investigated the spatial–temporal variations in rainwater chemistry characteristics in a rapidly urbanizing area of Shenzhen from 2007 to 2022. The mean pH during 2007–2022 was 5.12 ± 0.45, significantly higher than the 4.51 recorded in 2006. The electrical conductivity (EC) of rainfall in Shenzhen was 15.79 ± 2.63 μS/cm and showed a progressive decrease over the years. Human activities influenced the trends of SO_4^{2-}, which decreased, while NO_3^- and Cl^- increased over time. The cations Ca^{2+} and NH_4^+ decreased, and Mg^{2+}, Na^+, and K^+ increased. In recent years, Na^+ and Cl^-, typical sea-salt ions, became dominant with a continually rising contribution. The study also revealed that Shenzhen has been affected by both sulfuric and nitric acid rain, formally ushering in the nitric acid-type era. Spatial variations in rainfall chemicals were observed, especially for pH and conductivity, attributed to urban growth, distance from the coast, and industrial development. Moreover, the total ion concentration of rainwater demonstrated high values in the spring and low values in the summer due to the significant effect of precipitation.

Keywords: rainwater chemistry; Shenzhen; spatial–temporal variations; enrichment factor; post-industrial age

1. Introduction

Chemicals present in rainwater serve as indicators of the degree and state of air pollution [1], offering valuable insights into atmospheric composition changes and human impacts on the atmosphere [2]. Different compounds are emitted into the atmosphere due to natural processes and human activities [3]. Rainfall serves as the most efficient scavenging factor for removing particles or gaseous contaminants from the atmosphere [4–6]. However, excessive and prolonged atmospheric deposition can result in the eutrophication of the earth-surface hydrosphere [7], such as rivers, lakes, and reservoirs [8,9], further reducing the biodiversity of the aquatic ecosystems [10–12]. Studying chemical constituents in rainwater can aid in comprehending the origin, characteristics, concentration, and process for clearing atmospheric pollutants [13].

China's economy experienced rapid growth following the implementation of the reform and opening-up policy [14]. However, the heavy reliance on fossil fuels caused increased emissions of sulfur oxides (SO_x) and nitrogen oxides (NO_x) [1]. Due to the massive fossil fuel consumption, East Asia, particularly China, has become the third pole of SO_x and NO_x sources after Europe [13,15] and the United States [16]. Since 1990, acid precipitation has become a growing issue in China [2], which was previously considered one of the most critical regions for emissions of precursor gas of acid rain, with Northeast America and Central Europe following closely [17,18]. According to a 2008 investigation, the area affected by acid rain in China decreased to 13.8% of the overall land area [19].

However, the area of heavy acid rain has remained at a relatively high level compared to previous years when approximately 30% of the land area in China was afflicted by acid rain [20]. Due to China's severe acid rain problem, researchers have studied the forms, chemical composition characteristics, spatial–temporal distribution, and acidification mechanism of precipitation in urban and rural areas [1,2,21,22]. The Pearl River Delta and the Yangtze River Delta have been particularly focused on [23,24], as these regions are characterized by intense human activities, industrial production, motor vehicle emissions, and waste incineration, resulting in severe air pollution and acid deposition [21]. However, since the implementation of China's Air Pollution Action Plan, some new trends have emerged in the atmospheric environment [21]. For instance, China's emissions of NO_x decreased by 17% and SO_2 by 62% between 2010 and 2017 [25]. Li et al. (2016), Xu et al. (2020), and Ge et al. (2021) [26–28] also reported that the increase in pH value in rainwater is mainly attributed to the stricter control measures for SO_2 emissions implemented since the 13th Five Year Plan (2016–2020).

This study aims to investigate the emerging trend of rainwater chemistry characteristics, analyze the spatial and temporal variations in the chemical composition of rainwater, and identify the sources of the primary ion content in precipitation over a 16-year period in the densely populated area of Shenzhen city, China.

2. Study Area

The study area is located in the southern region of Guangdong Province, China, Shenzhen (22°26′–22°51′ N, 113°45′–114°37′ E), which is close to Hong Kong (Figure 1). Shenzhen's gross area is estimated to be 1997 km^2, based on a thorough assessment of land resources that was finished at the end of 2022 (including all islands) (http://pnr.sz.gov.cn/, accessed on 6 May 2023). Its terrain features a variety of landforms, with hills being the predominant feature followed by plains. The topography is higher in the Southeast and lower in the Northwest. The region experiences a southern subtropical monsoon climate, with a mean annual temperature of 22.4 °C (22–26 °C in spring and winter and 24–31 °C in summer and autumn) [14]. The rainy season spans from April to September (accounting for 80% of the annual rainfall), with an average yearly rainfall of 1925 mm. The dominant wind direction throughout the year is southeast wind. Southeast wind prevails in summer and northeast monsoon prevails in winter. During the summer, typhoons frequently hit the region.

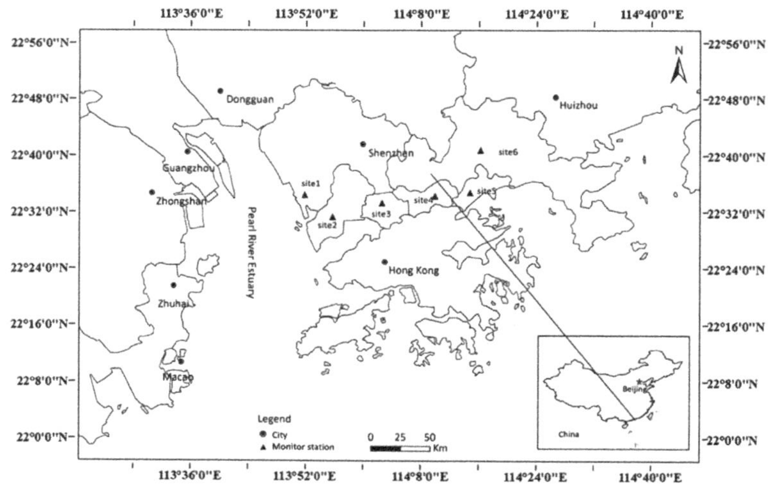

Figure 1. The location of Shenzhen city and monitor stations.

When Shenzhen was founded in 1979, it was primarily an agricultural region. Nonetheless, with the establishment of the Shenzhen Special Economic District (SED) in 1980, the city underwent rapid industrialization and urbanization processes, converting it into a newly developed city within two decades. This development was associated with environmental degradation, especially for the atmospheric environment. By the end of 2022, Shenzhen had a population of 17.66 million permanent residents, a Gross Domestic Product (GDP) of 463 billion (USD), a built-up area of 962 km^2, 3.95 million motor vehicles, and 2.6 million enterprises.

3. Methodology

3.1. Data Collection

Historical data on rainfall and the chemical composition of rainwater (annual or monthly averaged data) from six monitoring sites during the period 2007–2022 were collected from the Shenzhen Eco-environment Bureau (http://meeb.sz.gov.cn, accessed on 3 March 2023) (Figure 2). Over the 16-year period, around 9203 samples have been accumulated.

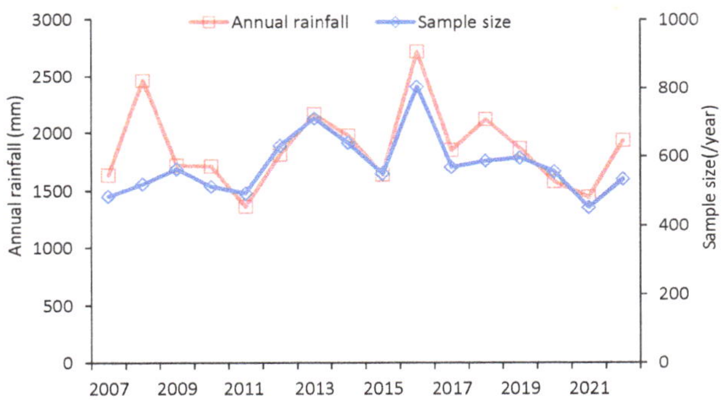

Figure 2. Annual rainfall and sampling number at Shenzhen during 2007–2022.

3.2. Method for Sample Collection and Treatment

This study includes a total of six monitoring sites, namely, Bao'an (site 1), Nanshan (site 2), Futian (site 3), Luohu (site 4), Yantian (site 5), and Longgang (site 6) (Figure 1). Land use in Bao'an, Nanshan, and Longgang is predominantly industrial; Futian and Luohu districts are primarily commercial and residential; Yantian district is focused on tourists. At all six sites, typical rain gauges (APS-2A, Biaozou Company, Shanghai, China; with a collector diameter of 300 mm and rain collection bottle volume of 5 L) were applied to collect rainwater samples. The collectors were positioned approximately 10 m above the ground, and the potential industrial plants were far from the rain gauges, more than 5 km, with no surrounding obstacles or obvious pollution sources near the sampling sites [26].

Before and after each collecting operation, the funnel and collector bottles were cleaned with triple-distilled water 3 times. In order to prevent dry deposition of gaseous and particulate pollutants, the collector was covered with a lid, which was opened only shortly before each rain event and sampling [24,26]. Continuous rainfall (24 h) was treated as one sample of precipitation. Prior to chemical analysis, the samples were filtered with a 0.45 μm filter membrane and kept in pre-cleaned polyethylene vials in a refrigerator at 4 °C.

The pH values of rainwater were determined in a laboratory setting using glass electrodes at 25 °C, and the electric conductivity was measured using a conductivity meter (Cond 315 i, WTW, Munich, Germany). All nine ions (K^+, Ca^{2+}, Na^+, Mg^{2+}, NH_4^+, Cl^-, F^-, NO_3^-, and SO_4^{2-}) were measured using Ion Chromatography (ILC-1, Waters, Milford, Massachusetts, USA) for all 6 sites during 2007–2022.

3.3. Quality Control Methods

The duplicate, blank, and standard reference material methods were used to control the quality. Each batch of rainwater underwent chemical composition analysis, which included at least two full program blanks and two laboratory blanks [29]. The ion content of blank samples was lower than the detection limit of the related analysis method. To further ensure the accuracy of the results, 10% of the samples from each batch were taken for parallel analysis, and the standard deviation of parallel samples was less than 10%. Additionally, a mixed standard solution with a concentration in the middle of the curve was taken as the quality control sample and was measured every 10 samples to check the stability of the instrument operation and ensure the accuracy of the test results [29].

In order to control the quality, data points outside of the range of (m − 3σ, m + 3σ), where m is the mean and σ is the standard deviation, were disregarded [30]. Moreover, this investigation did not consider precipitation volumes less than 1 mm, samples tainted with bird droppings or dust, or ion balances outside of ±20% [14]. According to the aforementioned criteria, approximately 12% of the samples were rejected.

3.4. Statistical and Calculated Methods

To determine the non-sea-salt (NSS) fractions for Cl^-, SO_4^{2-}, K^+, and Ca^{2+} concentrations, [Na^+] was used as a conservative tracer for sea salt contribution [31]. The sea salt contribution was then computed using accepted ionic ratios in seawater [32].

$$\mathrm{NSS} = X_p - C_p[x_r/C_r] \tag{1}$$

where C represents the concentration of the reference ion, and X represents the concentration of the ion of interest. Ions in precipitation and reference material are denoted by the subscripts p and r, respectively.

In this study, Ca^{2+} and NH_4^+ were considered the primary neutralizing components of rain (neutralization potential, *NP*), while SO_4^{2-} and NO_3^- served as the primary acidifying components of rain (acidic potential, *AP*) [14]. The equilibrium between acidity and alkalinity was evaluated using NP/AP. The NP to AP ratio is calculated as

$$NP/AP = ([\mathrm{nssCa}^{2+}] + [\mathrm{NH_4}^+])/([\mathrm{nssSO_4}^{2-}] + [\mathrm{NO_3}^-]) \tag{2}$$

The enrichment factor (*EF*) values for rainfall compositions were computed using Ca and Na as reference elements for continental and marine origin to estimate the terrestrial and marine contributions to rainwater [30,33]:

$$EF_{\mathrm{seawater}} = [X/\mathrm{Na}^+]_{\mathrm{rainwater}}/[X/\mathrm{Na}^+]_{\mathrm{seawater}} \tag{3}$$

$$EF_{\mathrm{soil}} = [X/\mathrm{Ca}^{2+}]_{\mathrm{rainwater}}/[X/\mathrm{Ca}^{2+}]_{\mathrm{soil}} \tag{4}$$

where X is the concentration of the target ion, and X/Na^+ of seawater is the ratio based on the composition of seawater [34]. In addition, X/Ca^{2+} of soil is the ratio based on the composition of crustal material [34].

Analysis of variance (ANOVA) was used to test the difference in ions among the different sites in Shenzhen. Comparisons were made using the least significant difference (LSD) method with $p < 0.05$.

4. Results and Discussions

4.1. pH, Electrical Conductivity, and Major Ions

From 2007 to 2022, the pH value of the precipitation ranged from 4.58 to 6.13, showing an upward trend ($r = 0.89$, $p < 0.001$) (Figure 3). The average pH value reached 5.12 ± 0.45, a noticeable increase from 4.51 in 2006, indicating a gradual weakening in the intensity of acid rain. The findings of this study are slightly higher than Zhang et al. (2007) and Jiang et al. (2019), who reported a pH value of 4.98 and 5.01, respectively, for rainfall in

Shenzhen [29,34]. The findings indicate a new trend in our study area after the implementation of China's Air Pollution Action Plan and the fight for "blue sky and clean water" in Shenzhen during the last decade [21]. Due to the difference in acidic component, the pH value of rainwater in this study is also higher than that of the background rainfall in Tokyo (4.7) reported by Uchiyama et al. (2017) [35], and it is comparable to that of Shanghai city (5.3) in South China [36] (Table 1). However, it is lower than that of Beijing city (6.8) in North China, Fuzhou city (6.0), Guiyang city (6.1), and Kunming city (6.6) in South China [1,22,34,37]. Because of the disparity in neutralizing capacity, our pH value was comparatively lower than that of several typical cities or regions throughout the world, such as East Malaysia (5.27), Costa Rica (5.90), and Sicily of Italy (6.7) [6,17,38,39].

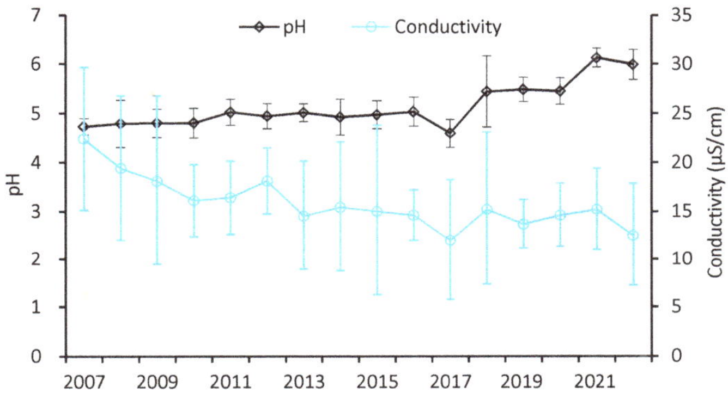

Figure 3. The pH value (mean ± SD) and conductivity (mean ± SD) in Shenzhen during 2007–2022.

Table 1. Rainfall chemicals at various sites in the world (μeq/L).

Area	Period	pH	SO_4^{2-}	NO_3^-	F^-	Cl^-	NH_4^+	Ca^{2+}	Na^+	Mg^{2+}	K^+
This study	2007–2022	5.12	34.45	23.9	1.98	29.53	25.49	29.88	24.92	6.75	5.24
Beijing, China [37]	2021	6.8	55.4	49.1	-	76.8	62.7	152.4	49.3	42.3	38
Guiyang, China [1]	2022–2023	6.1	197.3	93.2	20.6	29.5	205.0	209.0	19.7	29.0	20.4
Kunming, China [22]	2022–2023	6.6	104.8	54.9	15.9	22.4	52.1	465.4	18.0	30.3	18.3
Shanghai, China [36]	2016	5.3	59.9	41.1	3.8	19.2	68.2	32.4	15.6	7.1	3.2
Fuzhou, China [34]	2022–2023	6.0	55.1	86.6	4.2	64.0	76.3	145.3	52.5	21.1	10.7
Tokyo, Japan [35]	2012–2016	4.7	35.5	29.4	-	45.1	32.1	28.8	28.8	8.4	1.7
East Malaysia [6]	2017–2019	5.27	73.3	39.0	5.3	343.4	53.9	109.0	433.3	82.5	35.9
Costa Rica [38]	2020	5.90	29.9	11.0	-	24.5	18.9	44.6	12.9	16.8	13.3
Sicily, Italy [39]	2018–2019	6.7	80	14	7	470	-	130	350	90	9
Tatra, Poland [17]	2002–2019	5.95	89	30	1	20	65	56	12	19	6
Hawai'i, USA [40]	2018–2021	-	25.6	-	-	296.6	-	121.5	260.4	56.7	13.1

Notes: [1]. Zeng et al., 2004a [1]; [6]. Abdullah et al., 2022 [6]; [17]. Małecki et al., 2022 [17]; [22]. Zeng et al., 2024b [22]; [34]. Zeng et al., 2024c [34]; [35]. Uchiyama et al., 2017 [35]; [36]. Meng et al., 2019 [36]; [37]. Xu et al., 2020 [37]; [38]. Esquivel-Hernández et al., 2023 [38]; [39]. Brugnone et al., 2023 [39]; [40]. Brennis et al., 2023 [40].

The electrical conductivity (EC) of precipitation was 15.79 ± 2.63 μS/cm, showing an overall downward trend (r = −0.90, p = 0.001) from 2007 to 2022. This value was considerably lower than that of Beijing (58.2 μS/cm), reflecting the significantly better air quality in Shenzhen compared to Beijing [26].

Between 2007 and 2022, the averaged concentrations of SO_4^{2-} (34.35 ± 11.68 μeq/L) and NO_3^- (23.90 ± 8.82 μeq/L) were higher than those of Costa Rica and Hawai'i, nearly as high as Tokyo, but lower than those of Beijing, Guiyang, Kunming, Shanghai, and Fuzhou in China and East Malaysia, and Tatra in Poland [1,6,17,22,34–40]. The concentration of marine elements Cl^- (29.53 ± 10.82 μeq/L) and Na^+ (24.92 ± 10.66 μeq/L) in Shenzhen was significantly lower than that of East Malaysia, Sicily in Italy, Hawai'i in the USA, and Fuzhou in China but higher than that of Shanghai and Tokyo [1,6,17,22,34–40]. Because there was less agricultural activity, the amount of NH_4^+ (25.49 ± 8.74 μeq/L) in the rainwater was significantly lower than that in most other cities worldwide [1,6,17,22,34–36]. Ca^{2+} (29.88 ± 17.41 μeq/L) in rainwater was also much lower than that in most typical cities [1,6,17,22,34,36–38,40] due to the reduced anthropogenic activities in Shenzhen.

During the period 2007–2022, there were annual changes in the average concentrations (volume-weighted) of different anions in rainwater in Shenzhen (Figure 4). Due to the notable impact of SO_2 emission from industry and desulfurization of coal-fired power plants, SO_4^{2-} has exhibited a pronounced decreasing trend ($r = -0.90$, $p = 0.001$). Contrarily, NO_3^- showed an increasing tendency ($r = 0.54$, $p = 0.28$) over time, which has proven increasingly important in acid rain control. NO_x is one of the main sources of pollutants from motor vehicles in most cities in China [14]. In Shenzhen, the number of vehicles has proliferated over the past 20 years, reaching 1 million in 2006, 3 million in 2014, and 3.93 million in 2022. Since 2015, the government has begun restricting the issuance of new license plates, causing an accelerated growth of these plates. Another policy was to promote new energy vehicles, which now constitute approximately 10% of Shenzhen's car-washing. Regarding other anions, F^- exhibited a declining tendency ($r = -0.17$, $p = 0.96$), while Cl^- appeared to be growing ($r = 0.75$, $p = 0.02$).

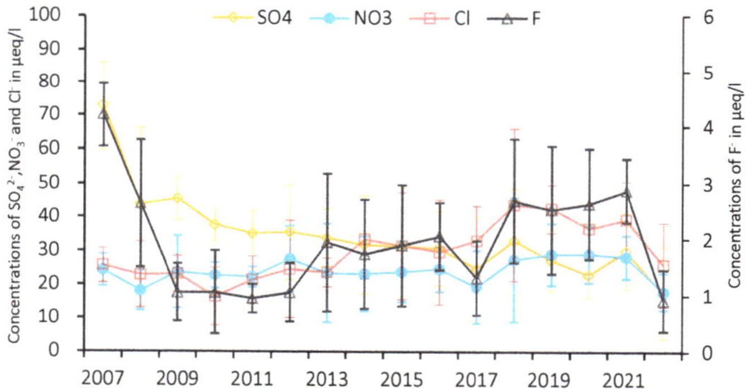

Figure 4. The variation in major anions (mean ± SD) in Shenzhen rainwater from 2007 to 2022.

With respect to the cations, Ca^{2+} demonstrated a distinct negative trend (r = −0.63, $p = 0.03$), predominantly related to increased urbanization and severely reduced constructive activities (Figure 5). Due to decreased farmland and agriculture activities, NH_4^+ displayed a decreasing tendency ($r = -0.44$, $p = 0.47$). With regard to the other cations, Mg^{2+}, Na^+, and K^+ all displayed an increased trend ($r = 0.34$, $p = 0.66$; $r = 0.80$, $p = 0.007$; $r = 0.71$, $p = 0.049$, respectively), indicating that the influence of the sea source on the chemical composition of rainfall has been growing in the last two decades.

The results of this study are similar to the long-term trend in some developed countries or regions [18,40]. For example, studies on acid rain revealed that, in their early stages, the rainwater's acidity substantially increased along with sulfate and nitrate concentrations [18,41]. The concentration of sulfates and the acidity of rainwater have thus significantly decreased throughout Europe and the United States after the 1980s [13,42].

In recent decades, both the neutralization potential (NP) and acidic potential (AP) have shown considerable declines ($r = -0.77$, $p = 0.015$; $r = -0.88$, $p = 0.001$), indicating a progressive reduction in crustal components and human factors (Figure 6). These findings led to an average *NP/AP* value of 1.02 ± 0.13 with a slightly increasing trend ($r = 0.53$,

$p = 0.29$) between 2007 and 2022, which was lower than Huang et al. (2008)'s former study [14].

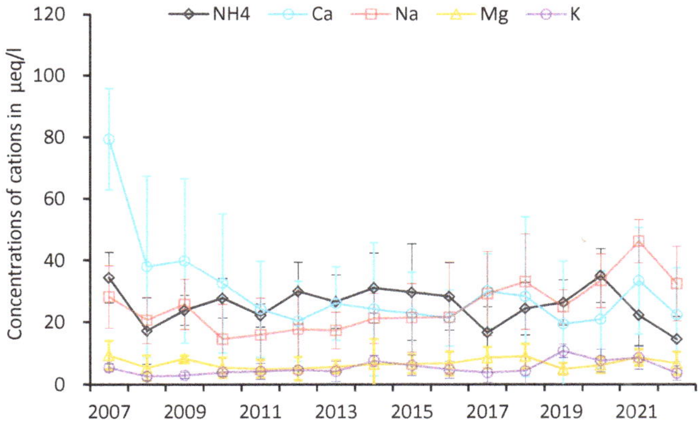

Figure 5. The changes in major cations (mean ± SD) in Shenzhen rainfall from 2007 to 2022.

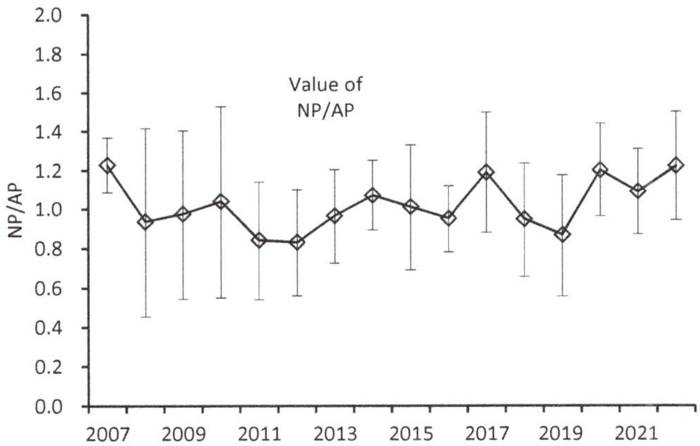

Figure 6. The value of neutralization potential (NP)/acidic potential (AP) (mean ± SD) of rainfall in Shenzhen during 2007–2022.

The average ratio $nssSO_4^{2-}/NO_3^-$ decreased from 2.73 to 0.92, showing a clear negative trend ($r = -0.91$, $p = 0.001$) throughout the study period (Figure 7). The ratio $[NO_3^-]/([nssSO_4^{2-}]+[NO_3^-])$ was used to determine the contribution of NO_3^- to acidification [30]. According to this study, the global contributions of atmospheric precipitation were between 60 and 70%, whereas the average amount of H_2SO_4 ranged from 36% to 73%, with a median of 52%, indicating a smaller contribution of sulfuric acid. Regarding HNO_3, the findings in Shenzhen were higher than the global estimates (30–40%), ranging from 27% to 64%, with an average of 48% [43]. This suggests that Shenzhen has entered an era dominated by nitric acid-type acid rain in recent years, after years of coexisting sulfuric acid and nitric acid-type acid rain. This is similar to the trend of other metropolises such as Tokyo (71%) and London (85%–99%) [13,35].

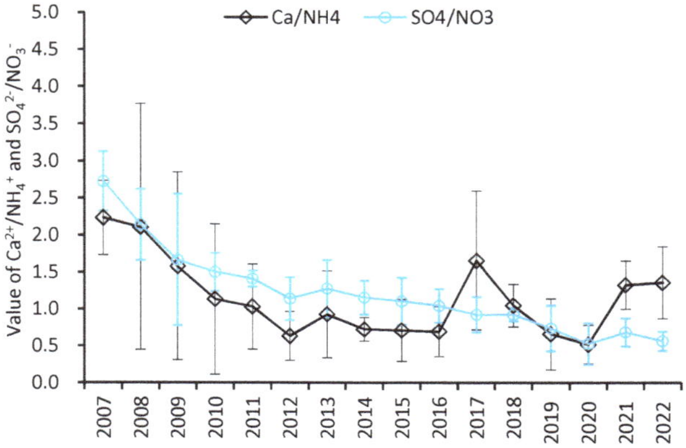

Figure 7. The value of Ca^{2+}/NH_4^+ and SO_4^{2-}/NO_3^- (mean ± SD) of rainfall in Shenzhen during 2007–2022.

The average ratio $nssCa^{2+}/[NH_4^+]$ decreased from 2.73 to 0.52, showing an obvious decreasing trend ($r = -0.96$, $p < 0.001$). Utilizing the ratio of $nssCa^{2+}/([nssCa^{2+}] + [NH_4^+])$ [30], the relative contribution of $nssCa^{2+}$ to neutralization was calculated. The averaged contribution of $nssCa^{2+}$ was found (from 36% to 80%, averaging 52%) to be slightly higher than that of NH_4^+ (from 20% to 64%, averaging 48%), suggesting that the neutralization abilities of calcium and ammonia were similar.

In the rapid urbanization stage, Ca^{2+} and SO_4^{2-} dominated the chemical composition of precipitation [14]. However, in the later stages of urbanization, the city gradually enters a soil- and marine-source dominant stage, and the influence of NO_3^- becomes more significant than SO_4^{2-} [13]. China's economic and social development plan generally has a milestone period of 5 years. The periods of 2007–2010 and 2011–2015 belong to rapid development stages, and with the implementation of China's Air Pollution Action Plan in 2016, Shenzhen has entered into the sustainable development stage of the economy and environment. In the present study, during the period 2007–2010, as a rapid urbanization state, SO_4^{2-} demonstrated the highest contribution (25.3%), followed by Ca^{2+} (21.2%), Na^+ (12.7%), NH_4^+ (12.5%), Cl^- (10.9%), and NO_3^- (10.8%) (Figure 8A). Additionally, we found in Figure 8B that SO_4^{2-} contributed the most (20.0%), followed by NH_4^+ (16.7%), Cl^- (16.1%), NO_3^- (14.4%), Ca^{2+} (14.17%), and Na^+ (11.2%). However, during 2016–2020, the dominant ions shifted from SO_4^{2-} (14.9%) to Cl^- (20.0%) (Figure 8C). The other ions, such as Na^+ (15.4%), NH_4^+ (14.2%), NO_3^- (14.0%), and Ca^{2+} (13.0%), also have important contributions during the period from 2016 to 2020. Meanwhile, during 2021–2022, we also found the dominant ions shifted from Cl^- (18.0%) to Na^+ (21.8%) (Figure 8D). The other ions, such as ca^{2+} (15.5%), SO_4^{2-} (13.1%), NO_3^- (12.8%), and NH_4^+ (10.2%), also have important contributions during the period from 2021 to 2022. In recent years, the contributions of Na^+ and Cl^- (sea-salt ions), increased gradually and became the dominant ions and were associated with the decreasing of crustal and anthropogenic sources due to the city gradually entering a post-industrial age. The typical ion compositions of Cl^-, Na^+, SO_4^{2-}, Mg^{2+}, Ca^{2+}, and K^+ from the sea salt near Shenzhen are 266.22 μeq/L, 229.5 μeq/L, 57.4 μeq/L, 52.1 μeq/L, 10.0 μeq/L, and 5.04 μeq/L, respectively [33]. Corral et al. (2020) also reported that volume-weighted concentrations of wet deposition were consistently predominated by sea salt in developed areas of the Southeast United States [44].

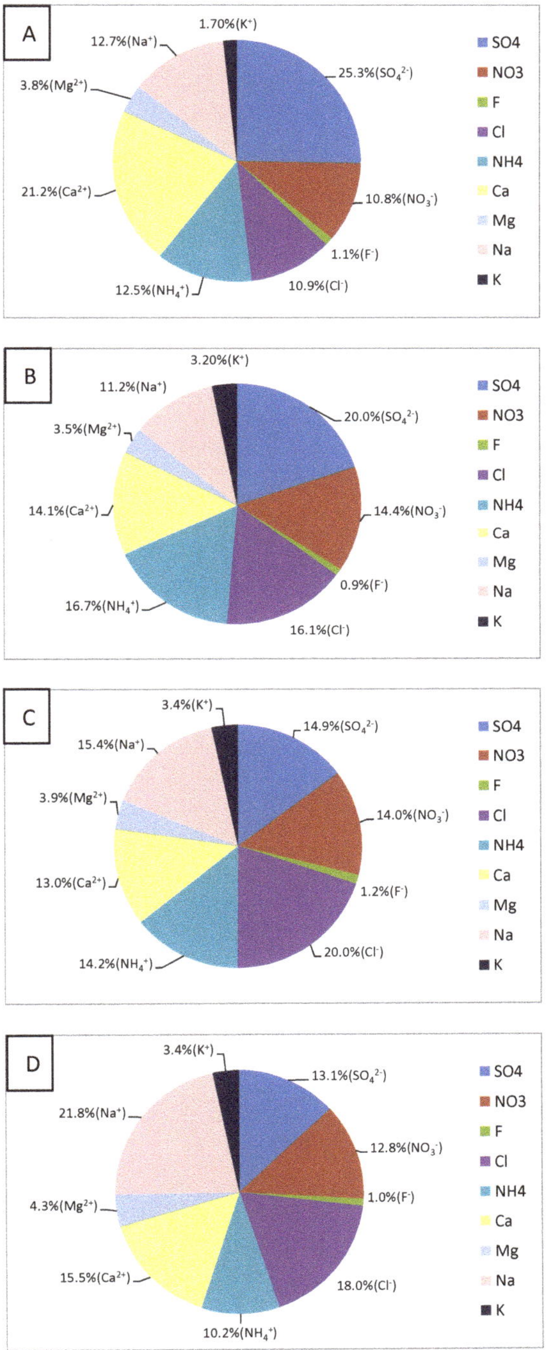

Figure 8. Chemical composition of rainwater during different stages: (**A**) (2007–2010), (**B**) (2011–2015), (**C**) (2016–2020), and (**D**) (2021–2022) in Shenzhen City.

According to previous studies, the rise in nitrogen oxidants has led to a decline in the NH_4^+/NO_3^- ratio, dropping from 5 to 2 in China over the past decade [24,45]. In our study, the NH_4^+/NO_3^- ratio declined from 1.41 to 0.86, averaging 1.06 ± 0.19 between 2007 and 2022 (Figure 9). According to the findings of Brimblecombe (2022) in Hong Kong and Jiang et al. (2019) in Shenzhen [29,46], this trend reflects a more rapid increase in NO_x emissions from industrial and traffic sources and a decrease in NH_4^+ emissions from agriculture sources. The low NH_4^+/NO_3^- ratio is consistent with the findings of Liu et al. (2013) and Liu et al. (2017) in nitrate-dominant coastal cities of China [45,47], such as Qingdao, Daya Bay, and Fuzhou in China [34,48,49].

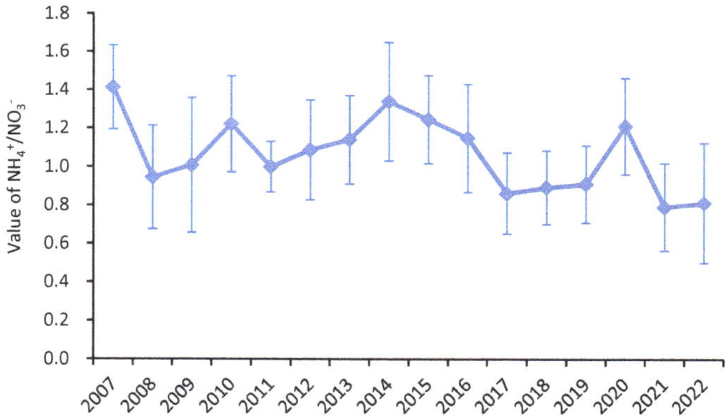

Figure 9. The value (mean ± SD) of NH_4^+/NO_3^- of rainfall in Shenzhen during 2007–2022.

4.2. EF and Source Determination

A variety of artificial and natural variables affect the chemical composition of rainwater [50], which is largely determined by local sources, although the participation of a continental airmass was evident for some systems [1,22,34]. In this investigation, the enrichment factor (EF) was applied to differentiate between aerosol sources. An EF value significantly below 1 or substantially above 1 is regarded as diluting or enriching the reference source, respectively [33]. SO_4^{2-}, K^+, Mg^{2+}, Cl^-, and NO_3^- had average EF values of 11.5, 1.17, 0.52, 318.88, and 380.89, respectively, in soil sources (Table 2). In contrast, the mean EF values for SO_4^{2-}, Ca^{2+}, K^+, Mg^{2+}, and Cl^- originating from sea salt were 5.52, 27.27, 9.57, 1.19, and 1.02, respectively. Consequently, SO_4^{2-} predominantly stemmed from anthropogenic origins, albeit crustal and sea sources also exerted significant influence. Cl^- was especially from sea-salt sources, and NO_3^- largely derived from anthropogenic sources. Sea salt served as the primary source of K^+ and Mg^{2+}.

Table 2. EF for soil and sea-salt sources.

Index	SO_4^{2-}	Ca^{2+}	K^+	Mg^{2+}	Cl^-	NO_3^-
Soil ratio	0.1	1	0.15	0.43	0.0031	0.0021
Rainwater ratio	1.15	1	0.18	0.23	0.99	0.80
EF soil	11.50	1	1.17	0.52	318.88	380.89
	SO_4^{2-}	Ca^{2+}	K^+	Mg^{2+}	Cl^-	NO_3^-
Seawater ratio	0.25	0.044	0.022	0.227	1.16	-
Rainwater ratio	1.38	1.20	0.21	0.27	1.19	-
EF sea salt	5.52	27.27	9.57	1.19	1.02	-

The main source of SO_4^{2-} was anthropogenic (75.2%), followed by oceanic (16.7%) and crust (8.1%) sources (Table 3). An anthropogenic origin dominated the sources of NO_3^- (99.8%), and only 0.2% came from crust sources. Cl^- primarily came from sea-salt sources (97.8%), followed by anthropogenic and soil sources (1.9% and 0.3%, respectively). Ca^{2+} and Mg^{2+} primarily came from soil sources (92.7% and 83.4%, respectively): however, the sea source (79.3%) was the dominant contribution of K^+, with the second being the soil source (20.7%). The results are consistent with the study conducted by Zeng et al. (2022) in the South Chinese coastal city of Fuzhou [51].

Table 3. Contributions from sources for various ionic compositions in rainfall.

Ion	Sea Salt Fraction (%)	Terrestrial Fraction	
		Crust Fraction (%)	Anthropogenic Source (%)
SO_4^{2-}	16.7	8.1	75.2
NO_3^-	-	0.2	99.8
Cl^-	97.8	0.3	1.9
Ca^{2+}	7.3	92.7	-
Mg^{2+}	16.6	83.4	-
K^+	79.3	20.7	-

4.3. Seasonal Variation

The total ion concentrations of precipitation in Shenzhen presented an obvious variation on a monthly scale (Figure 10), with higher concentrations in spring and lower concentrations in summer. Meanwhile, the ion concentrations from anthropogenic, marine, and crust sources were higher in spring and lowest in summer. Besides the seasonal difference in source emissions, precipitation was a primary factor. Relevant studies indicated that increased precipitation caused the ion concentration to decline through diffusion, evaporation, and dilution [7]. Statistical analysis showed that the concentrations of ions except H^+ were negatively correlated with precipitation ($r = -0.24 \sim -0.67$), reflecting the removal of atmospheric particles by precipitation. Shenzhen experiences concentrated rainfall mostly in the summer and autumn, with little precipitation in spring and winter. Although there is less rainfall in winter, the convection and diffusion conditions are superior compared to other seasons, which may explain the seasonal distribution of precipitation ion content [33]. This finding is consistent with the report of Zhang et al. (2020) in the Guiling city of Guangxi Province, China [21], and Brugnone et al. (2023) in the industrial areas of Milazzo and Priolo Gargallo, Italy, which found seasonal atmospheric deposition followed the concentrations of various ion species that also depend on rainfall amounts [38]. Khan and Sarwar (2014) also reported that heavy rains washed off TSP and suppressed the lifting of dust particles, leading to a drop in Ca and Mg concentrations in Pakistan [52].

According to the study reported by Yu et al. (2015), the total ion concentration on precipitation cleanliness can be categorized into four levels of pollution [23]. These categories are light pollution, which ranges from 100 to 300 μeq/L, moderate pollution, which is between 300 and 500μeq/L, severe pollution, which is between 500 and 1000 μeq/L, and extremely heavy pollution, which is above 1000 μeq/L. Shenzhen's rainwater was classified under moderate pollution from January to April and light pollution in other months.

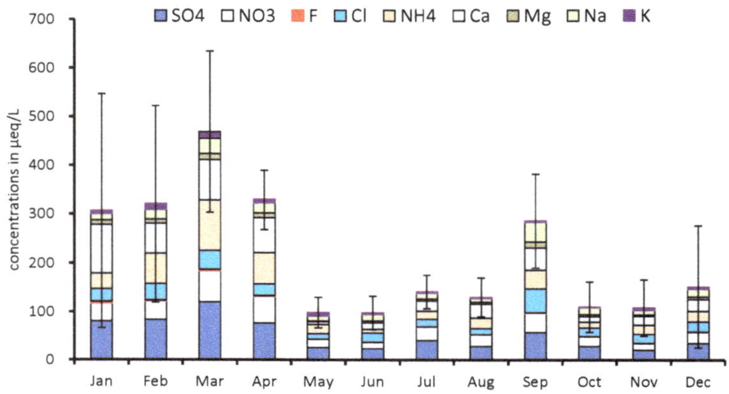

Figure 10. Seasonal variation in rainwater chemicals in Shenzhen (error bar is for the total ion concentrations).

4.4. Spatial Variation

The chemical composition of rainwater in Shenzhen exhibited variations attributable to rapid urban development, distance from the ocean, and industrial growth. Among the six districts, Luohu, Futian, and Nanshan have a relatively high degree of urbanization. Nanshan, Bao'an, and Longgang have developed industries, while Nanshan, Bao'an, and Yantian are close to the sea. In general, the highest rainwater concentrations were observed at the Nanshan site (264 μeq/L) and Bao'an site (261 μeq/L), followed by the Luohu site (226 μeq/L) and Longgang site (222 μeq/L), with the Yantian site (213 μeq/L) and Futian site (183 μeq/L) registering lower concentrations (Figure 11). The spatial disparities among these sites were linked to human activities and urban expansion [17,26,39].

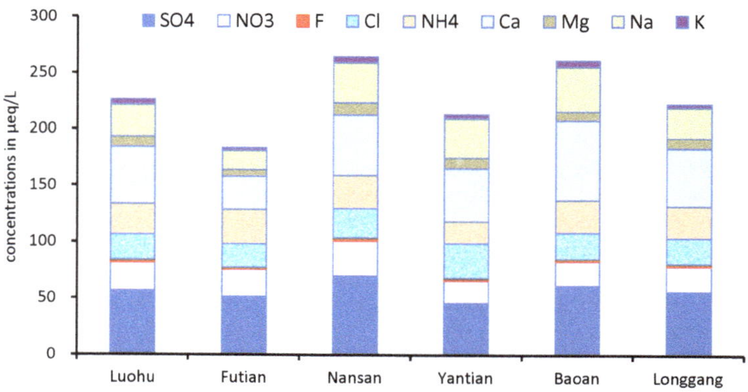

Figure 11. Spatial variations in rainwater chemicals of different sites in Shenzhen.

The study of ion differences unveiled significant regional variations in Ca^{2+} and K^{+} originating from the Earth's crust. In contrast, the differences between other ions from the ocean and anthropogenic sources were negligible. Conductivity and pH were also influenced by the distribution of ions across various regions. The pH value was lower in Futian and Yantian sites (4.50 and 4.57, respectively), followed by Nanshan and Longgang (4.83 and 4.90, respectively), and higher in Luohu and Bao'an (5.05 and 5.15, respectively) (Figure 12A). The variations in pH between the six sites were statistically significant ($p < 0.05$). Among the six sites, the conductivity was higher (33.2 μS/cm) at the Nanshan site, while it was lower at the other five sites (17.1~22.8 μS/cm) (Figure 12B).

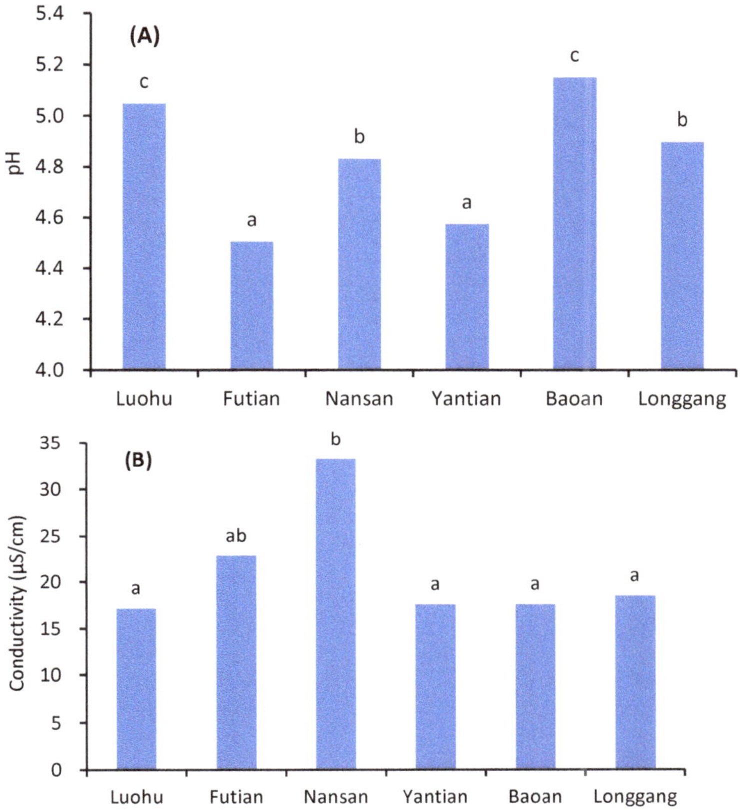

Figure 12. pH (**A**) and conductivity (**B**) of rainwater at different sites in Shenzhen (the letters on the bar chart indicate the different levels of ANOVA analysis).

5. Conclusions

The present study investigated the novel pattern of spatial–temporal fluctuations in rainwater composition in Shenzhen from 2007 to 2022. Our findings revealed that the mean pH value reached 5.12 ± 0.45 during the period 2007–2022, which was significantly higher than the value of 4.51 in 2006. The EC value of precipitation in Shenzhen was 15.79 ± 2.63 μS/cm and showed a gradual downward trend over the study period. Among the major anions, SO_4^{2-} showed a decreasing trend, while NO_3^- and Cl^- showed an increasing trend, due to intensive human impact. Thus, Shenzhen needs to pay more attention to the control of NOx. Among the cations, Ca^{2+} and NH_4^+ presented a decreasing trend, while Mg^{2+}, Na^+, and K^+ exhibited an increasing trend. Na^+ and Cl^-, typical components of sea salt, have recently increased and become the dominant ions accompanied by a decreasing contribution from crustal and anthropogenic sources as the city enters the post-industrial age. The study also found that, in recent years, the nitric acid type has replaced the former combined acid rain of sulfuric acid and the nitric acid type in Shenzhen. These findings have significant ramifications for future studies on the effects of acid deposition on ecosystem functioning, non-point pollution, and rainwater harvesting.

According to the EF analysis, anthropogenic sources accounted for the majority of SO_4^{2-} (75.2%), followed by ocean sources (16.7%) and crustal sources (8.1%). The majority (99.8%) of NO_3^-'s sources were anthropogenic, with only 0.2% coming from crust sources. In contrast, Cl^- was primarily from sea-salt sources (97.8%), followed by anthropogenic sources and soil sources (1.9% and 0.3%, respectively). Ca^{2+} and Mg^{2+} primarily came from soil sources (92.7% and 83.4%, respectively). K^+ was mainly from sea sources (79.3%), with the remaining 20.7% from soil sources.

Obvious spatial variations in rainfall chemicals were found in Shenzhen, especially for the pH value and conductivity, which were related to the differences in urban construction, distance from the ocean, and industrial development. The total ion concentration of rainwater in Shenzhen presented a sinusoidal variation on the monthly scale, with higher contributions in spring and lower contributions in summer, largely influenced by precipitation.

Author Contributions: Y.H.: Conceptualization; Funding acquisition; Project administration; Resources; Writing—original draft; Writing—review and editing; Supervision. J.Y.: Writing—review and editing; Validation; Visualization. C.L., R.J. and Q.L.: Data curation; Formal analysis; Investigation; Methodology; Software. All authors have read and agreed to the published version of the manuscript.

Funding: This study was supported by the Shenzhen Science and Technology Innovation Program (Grant NO. KCXST20221021111416039).

Institutional Review Board Statement: The study does not need ethical approval.

Informed Consent Statement: Not applicable.

Data Availability Statement: The authors declared that no external data had been used in this study.

Conflicts of Interest: Authors are employees of Shenzhen Water Planning & Design Institute Co., Ltd. The paper reflects the views of the scientists and not the company.

References

1. Zeng, J.; Han, G.L.; Wu, Q.X.; Peng, M.X.; Ge, X.; Mao, S.J.; Wang, Z.J.; Ma, Q. Chemical evolution of rainfall in China's fi eco-civilization demonstration city: Implication for the provenance identification of pollutants and rainwater acid neutraliza *Sci. Total Environ.* **2024**, *910*, 168567. [CrossRef] [PubMed]
2. Shi, X.F.; Niu, H.W.; He, Y.Q.; Qi, C.S.; Pu, T.; Shi, X.Y. Characteristics of rainwater chemistry in Lijiang-Yulong snow moun *Environ. Chem.* **2017**, *36*, 994–1002. (In Chinese with English Abstract)
3. Qu, L.L.; Xiao, H.Y.; Zheng, N.J.; Zhang, Z.Y.; Xu, Y. Comparison of four methods for spatial interpolation of estimat atmospheric nitrogen deposition in South China. *Environ. Sci. Pollut. Res.* **2017**, *24*, 2578–2588. [CrossRef] [PubMed]
4. Niu, H.; He, Y.; Lu, X.X.; Shen, J.; Du, J.; Zhang, T.; Pu, T.; Xin, H.; Chang, L. Chemical composition of rainwater in the Yul snow mountain region, southwestern china. *Atmos. Res.* **2014**, *144*, 195–206. [CrossRef]
5. Zhao, P.; Lu, W.; Hong, Y.; Chen, J.; Huang, Q. Long-term wet precipitation of pm2.5 disturbed the gut microbiome and inhib the growth of marine medaka oryzias melastigma. *Sci. the Total Environ.* **2021**, *755 Pt 1*, 142512. [CrossRef]
6. Abdullah, S.N.F.; Ismail, A.; Juahir, H.; Ahmad, R.B.; Lananan, F.; Hashim, N.M.; Ariffin, N.; Zali, M.A.; Mohd, T.A.T.; Huss M.H.F.; et al. Chemical composition of rainwater harvested in East Malaysia. *Environ. Eng. Res.* **2022**, *27*, 200508. [CrossRef]
7. Qin, H.P.; Tang, Q.L.; Wang, L.Y.; Fu, G.T. The impact of atmospheric wet deposition on roof runoff quality in an urbanized a *Hydrol. Res.* **2015**, *46*, 880–891. [CrossRef]
8. Wu, Q.; Han, G. δ13C DIC tracing of dissolved inorganic carbon sources at Three Gorges Reservoir, China. Water science a technology. *A J. Int. Assoc. Water Pollut. Res.* **2018**, *77*, 555–564.
9. Wang, Z.J.; Yue, F.J.; Wang, Y.C.; Qin, C.Q.; Ding, H.; Xue, L.L.; Li, S.L. The effect of heavy rainfall events on nitrogen patterns agricultural surface and underground streams and the implications for karst water quality protection. *Agric. Water Manag.* **20** *266*, 107600. [CrossRef]
10. Sharpley, A.; Wang, X.Y. Managing agricultural phosphorus for water quality: Lessons from the USA and China. *J. Environ.* **2014**, *26*, 1770–1782. [CrossRef]
11. Liang, T.; Tong, Y.L.; Wang, X.H.; Wang, L.Q. Release of reactive phosphorus from sediments in Dongting Lake linked with Yangtze River. *Environ. Chem.* **2017**, *14*, 48–54. [CrossRef]
12. Zhang, X.; Lin, C.Y.; Zhou, X.L.; Lei, K.; Guo, B.B.; Cao, Y.X.; Lu, S.; Liu, X.T.; He, M.C. Concentrations, fluxes, and potent sources of nitrogen and phosphorus species in atmospheric wet deposition of the Lake Qinghai Watershed, China. *Sci. To Environ.* **2019**, *682*, 523–531. [CrossRef] [PubMed]

Keresztesi, Á.; Birsan, M.V.; Nita, I.A.; Bodor, Z.; Szép, R. Assessing the neutralization, wet deposition and source contributions of the precipitation chemistry over Europe during 2000–2019. *Environ. Sci. Eur.* **2019**, *31*, 50. [CrossRef]
Huang, Y.L.; Wang, Y.L.; Zhang, L.P. Long-term trend of the chemical composition of wet atmospheric precipitation during 1986–2006 at Shenzhen City, China. *Atmos. Environ.* **2008**, *42*, 3740–3750. [CrossRef]
Martuzzi, M.; Pasetto, R.; Martin-Olmedo, P. Industrially Contaminated Sites and Health. *J. Environ. Public Health* **2014**, 198574. [CrossRef]
Vet, R.; Artz, R.S.; Carou, S.; Shaw, M.; Ro, C.-U.; Aas, W.; Baker, A.; Bowersox, V.C.; Dentener, F.; Galy-Lacaux, C.; et al. A global assessment of precipitation chemistry and deposition of sulfur, nitrogen, sea salt, base cations, organic acids, acidity and pH, and phosphorus. *Atmos. Environ.* **2014**, *93*, 3–100. [CrossRef]
Małecki, J.J.; Matyjasik, M.; Krogulec, E.; Porowska, D. Long-term trends and factors influencing rainwater chemistry in the Tatra Mountains, Poland. *Geol. Geophys. Environ.* **2022**, *48*, 19–38. [CrossRef]
Pan, D.; Mauzerall, D.L.; Wang, R.; Guo, X.H.; Puchalski, M.; Guo, Y.X.; Song, S.J.; Tong, D.; Sullivan, A.P.; Schichtel, B.A.; et al. A reduction in aerolized ammonium in the rural USA and increased ammonia deposition near emission hotspots. *Nat. Geosci.* **2024**, *17*, 594–595.
State Environmental Protection Administration of China. *China Environmental Statement of 2008*; State Environmental Protection Administration of China: Beijing, China, 2009. (In Chinese)
Xie, S.Y.; Wang, S.J.; Yu, Y.; Liu, L.; Zhang, F.Y. Analysis on acid rain status and its change trends in China from 2003 to 2018. *Environ. Monit. China* **2020**, *36*, 80–88. (In Chinese with English Abstract)
Zhang, Q.H.; Li, Y.S.; Yu, S.; Luo, Z.X. Characteristics and source analysis of chemical composition of atmospheric precipitation in Guilin City, Southwest China. *Environ. Chem.* **2020**, *39*, 229–239. (In Chinese with English Abstract)
Zeng, J.; Han, G.L.; Wu, Q.X.; Ma, Q.; Mao, S.J.; Ge, X.; Wang, Z.J.; Yang, C. Evolution and determinants of precipitation chemistry in a plateau city, China: Insight from the pollutants source identification and environmental effect. *Urban Clim.* **2024**, *55*, 101872. [CrossRef]
Yu, Y.J.; Cheng, Z.; Li, J.; Chao, W.J.; Sun, J.; Shao, X.Y. Analysis of chemical composition and source of precipitation in Changzhou in 2014. *Environ. Chem.* **2015**, *34*, 2136–2138. (In Chinese with English Abstract)
Wu, Y.; Zhang, J.; Liu, S.; Jiang, Z.; Huang, X. Aerosol concentrations and atmospheric dry deposition fluxes of nutrients over Daya Bay, South China Sea. *Mar. Pollut. Bull.* **2018**, *128*, 106–114. [CrossRef] [PubMed]
Zheng, B.; Tong, D.L.; Meng, L.; Fei, H.; Geng, C.P.; Li, G.N.; Li, H.Y.; Peng, X.; Qi, L.Q.; Yan, J.; et al. Trends in China's anthropogenic emissions since 2010 as the consequence of clean air actions. *Atmos. Chem. Phys.* **2018**, *18*, 14095–14111. [CrossRef]
Wu, T.; Cui, Y.; Lian, A.; Tian, Y.; Li, R.; Liu, X.; Yan, J.; Xue, Y.; Liu, H.; Wu, B. Vehicle emissions of primary air pollutants from 2009 to 2019 and projection for the 14th five-year plan period in Beijing, china. *J. Environ. Sci.* **2023**, *124*, 513–521. [CrossRef]
Ge, B.; Xu, D.; Wild, O.; Yao, X.; Wang, J.; Chen, X.; Tan, Q.; Pan, X.; Wang, Z. Inter-annual variations of wet deposition in Beijing from 2014–2017: Implications of below-cloud scavenging of inorganic aerosols. *Atmos. Chem. Phys.* **2021**, *21*, 9441–9454. [CrossRef]
Li, B.; Wang, F. Chemical characteristics of atmospheric precipitation in Qingpu district of Shanghai. *Environ. Monit. China* **2016**, *32*, 24–29.
Jiang, B.Y.; He, L.; Chen, D.H.; Ge, T.A.; Liang, Z.W.; Wang, M.J.; Wang, X.Y.; Li, S.A. Wet deposition flux and sources of heavy metals in precipitation in the coastal area of Shenzhen. *Environ. Chem.* **2019**, *38*, 1460–1473. (In Chinese with English Abstract)
Safai, P.D.; Rao, P.S.P.; Momin, G.A.; Ali, K.; Chate, D.M.; Praveen, P.S. Chemical composition of precipitation during 1984–2002 at Pune, India. *Atmos. Environ.* **2004**, *38*, 1705–1714. [CrossRef]
Laouali, D.; Galy-Lacaux, C.; Diop, B.; Delon, C.; Orange, D.; Lacaux, J.P.; Akpo, A.; Lavenu, F.; Gardrat, E.; Castera, P. Long term monitoring of the chemical composition of precipitation and wet deposition fluxes over three Sahelian savannas. *Atmos. Environ.* **2012**, *50*, 314–327. [CrossRef]
Akkoyunlu, B.O.; Tayanç, M. Analyses of wet and bulk deposition in four different regions of Istanbul, Turkey. *Atmos. Environ.* **2003**, *37*, 3571–3579. [CrossRef]
Zhang, M.Y.; Wang, S.J.; Wu, F.C.; Yuan, X.H.; Zhang, Y. Chemical compositions of wet precipitation and anthropogenic influences at a developing urban site in southeastern China. *Atmos. Res.* **2007**, *84*, 311–322. [CrossRef]
Zeng, J.; Han, G.L.; Wu, Q.X.; Qu, R.; Ma, Q.; Chen, J.W.; Mao, S.J.; Ge, X.; Wang, Z.J.; Ma, Z.H. Significant influence of urban human activities and marine input on rainwater chemistry in a coastal large city, China. *Water Res.* **2024**, *257*, 121657. [CrossRef] [PubMed]
Uchiyama, R.; Okochi, H.; Katsumi, N.; Ogata, H. The impact of air pollutants on rainwater chemistry during "urban-induced heavy rainfall" in downtown Tokyo, Japan. *J. Geophys. Res. Atmos.* **2017**, *122*, 6502–6519. [CrossRef]
Meng, Y.; Zhao, Y.; Li, R.; Li, J.; Cui, L.; Kong, L.; Fu, H. Characterization of inorganic ions in rainwater in the megacity of Shanghai: Spatiotemporal variations and source apportionment. *Atmos. Res.* **2019**, *222*, 12–24. [CrossRef]
Xu, W.; Wen, Z.; Shang, B.; Dore, A.J.; Tang, A.; Xia, X.; Zheng, A.; Han, M.; Zhang, L.; Zhao, Y.; et al. Precipitation chemistry and atmospheric nitrogen deposition at a rural site in Beijing, China. *Atmos. Environ.* **2020**, *223*, 117253. [CrossRef]
Esquivel-Hernández, G.; Sánchez-Murillo, R.; Villalobos-Córdoba, D.; Monteiro, L.R.; Villalobos-Forbes, M.; Sánchez-Gutiérrez, R.; Cotrim, M.E.B.; Matiatos, I. Exploring the acid neutralizing effect in rainwater collected at a tropical urban area: Central Valley, Costa Rica. *Atmos. Pollut. Res.* **2023**, *14*, 101845. [CrossRef]

39. Brugnone, F.; D'Alessandro, W.; Parello, F.; Liotta, M.; Bellomo, S.; Prano, V.; Li Vigni, L.; Sprovieri, M.; Calabrese, S. Atmosph Deposition around the Industrial Areas of Milazzo and Priolo Gargallo (Sicily–Italy)—Part A: Major Ions. *Int. J. Environ. . Public Health* **2023**, *20*, 3898. [CrossRef]
40. Brennis, T.; Lautze, N.; Whittier, R.; Torri, G.; Thomas, D. Understanding the origins of and influences on precipitation major chemistry on the Island of O'ahu, Hawai'i. *Environ. Monit. Assess.* **2023**, *195*, 1265. [CrossRef]
41. Weber, R.J.; Guo, H. High aerosol acidity despite declining atmospheric sulfate concentrations over the past 15 years. *Nat. Ge* **2016**, *9*, 282–285. [CrossRef]
42. Fornaro, A.; Gutz, I. Wet deposition and related atmospheric chemistry in the So Paulo metropolis, Brazil. Part 3: Trend precipitation chemistry during 1983–2003. *Atmos. Environ.* **2006**, *40*, 5893–5901. [CrossRef]
43. Migliavacca, D.; Teixeira, E.C.; Wiegand, F.; Machado, A.C.M.; Sanchez, J. Atmospheric precipitation and chemical composi of an urban site, Guaíba Hydrographic Basin, Brazil. *Atmos. Environ.* **2005**, *39*, 1829–1844. [CrossRef]
44. Corral, A.F.; Dadashazar, H.; Stahl, C.; Edwards, E.L.; Zuidema, P.; Sorooshian, A. Source Apportionment of Aerosol at a Coa Site and Relationships with Precipitation Chemistry: A Case Study over the Southeast United States. *Atmosphere* **2020**, *11*, 1 [CrossRef] [PubMed]
45. Liu, X.J.; Zhang, Y.; Han, W.X.; Tang, A.H.; Shen, J.L.; Cui, Z.L.; Vitousek, P.; Erisman, J.W.; Goulding, K.; Christie, P.; e Enhanced nitrogen deposition over China. *Nature* **2013**, *494*, 459–462. [CrossRef]
46. Brimblecombe, P. Trends in secondary inorganic particles in Hong Kong, 1995–2020. *Atmos. Environ.* **2022**, *268*, 118801. [Cross
47. Liu, X.; Xu, W.; Duan, L.; Du, E.; Pan, Y.; Lu, X.; Zhang, L.; Wu, Z.; Wang, X.; Zhang, Y.; et al. Atmospheric nitrogen emiss deposition, and air quality impacts in China: An overview. *Curr. Pollut. Rep.* **2017**, *3*, 65–77. [CrossRef]
48. Luo, X.S.; Tang, A.H.; Shi, K.; Wu, L.H.; Li, W.Q.; Shi, W.Q.; Shi, X.K.; Erisman, J.W.; Zhang, F.S.; Liu, X.J. Chinese coastal seas facing heavy atmospheric nitrogen deposition. *Environ. Res. Lett.* **2014**, *9*, 095007. [CrossRef]
49. Wu, Y.; Zhang, J.P.; Liu, S.L.; Jiang, Z.J.; Arbi, I.; Huang, X.P.; Macreadie, P.I. Nitrogen deposition in precipitation to a monsc affected eutrophic embayment: Fluxes, sources, and processes. *Atmos. Environ.* **2018**, *182*, 75–86. [CrossRef]
50. Xing, J.W.; Song, J.M.; Yuan, H.M.; Li, X.G.; Li, N.; Duan, L.Q.; Qu, B.X.; Wang, Q.D.; Kang, X.M. Chemical characteris deposition fluxes and source apportionment of precipitation components in the Jiaozhou Bay, North China. *Atmos. Res.* **2017**, 10–20. [CrossRef]
51. Zeng, J.; Han, G.L.; Zhang, S.T.; Xiao, X.H.; Li, Y.K.; Gao, X.; Wang, D.; Qu, R. Rainwater chemical evolution driven by extre rainfall in megacity: Implication for the urban air pollution source identification. *J. Clean. Prod.* **2022**, *372*, 133732. [CrossRef
52. Khan, M.N. Sarwar, A. Chemical composition of wet precipitation of air pollutants: A case study in Karachi. *Pak. Atmósfera* **2** *27*, 35–46. [CrossRef]

atmosphere

MDPI

icle

nalysis of Forest Fire Emissions and Meteorological Impacts in outhwestern China Based on Multi-Source atellite Observations

ıgli Fang, Yu Han, Junbo Lin and Wenkai Guo *

School of Environmental Science and Engineering, Southwest Jiaotong University, Chengdu 611756, China; fanglingli@my.swjtu.edu.cn (L.F.); hany@my.swjtu.edu.cn (Y.H.); 1286965833@my.swjtu.edu.cn (J.L.)
* Correspondence: guowk@swjtu.edu.cn; Tel.: +86-188-9371-7453

Abstract: Amid the growing frequency of forest fires in southwestern China, this study aims to quantify pollutant emissions and identify key meteorological drivers using multi-source satellite data. Active fire data from Himawari-8/9, MODIS, and VIIRS were integrated to construct a top-down emission inventory for 2016–2023, while the Geodetector method was applied to evaluate meteorological influences. Results indicate mean annual emissions ($\times 10^3$ t·a^{-1}) of 5623.58 ($\pm$1554.33) for CO_2, 356.84 ($\pm$98.63) for CO, and substantial amounts of particulate and gaseous pollutants. Spatially, Yunnan and Sichuan were the dominant emitters; temporally, emissions peaked in January–April and November–December, with daytime levels surpassing nighttime levels. Relative humidity was identified as the dominant meteorological driver (Q = 0.1223), while the interaction between temperature and relative humidity (Q = 0.1486) further enhanced explanatory power. These findings improve the precision of emission inventories and provide essential support for regional fire management and air quality modeling in complex environments.

Keywords: satellite remote sensing; multi-satellite fire detection; forest fire emissions; emission inventory; carbon dioxide; Geodetector

1. Introduction

Against the backdrop of global climate change, extreme weather events have become increasingly frequent, leading to a steady rise in both the frequency and intensity of forest fires. These fires have emerged as one of the major environmental issues worldwide [1]. As a highly destructive natural disaster, forest fires not only destroy large areas of forest vegetation but also disrupt ecosystem structure and function, cause habitat loss, and seriously threaten biodiversity [2]. Large quantities of smoke and particulate matter released during fire events are rich in black carbon (BC) and organic carbon (OC), making forest fires an important contributor to greenhouse gas and aerosol emissions. These emissions can further degrade air quality and pose substantial risks to human health [3–5].

Southwestern China, home to the nation's second-largest expanse of natural forests, possesses abundant forest resources but is highly susceptible to wildfires. Its dense vegetation, complex topography, and variable climate contribute to a persistently high fire incidence [6,7]. In recent years, the atmospheric effects of forest fires in this region have become increasingly evident, with CO_2 emissions accounting for approximately 59.23% of the total carbon released from biomass burning and exhibiting a distinct seasonal pattern [8]. This seasonal clustering is closely linked to regional climatic conditions, as meteorological factors strongly influence fire ignition and propagation. High temperatures, low relative humidity, and moderate wind speeds are widely recognized as dominant drivers of wildfire occurrence [9]. Elevated temperature increases fuel flammability and accelerates moisture loss, while lower humidity reduces resistance to desiccation, and limited precipitation exacerbates dryness, collectively raising ignition probability. Meanwhile, moderate winds not only enhance oxygen supply but also accelerate flame spread and expand the affected area [10]. Therefore, understanding the emission characteristics of forest fires in southwestern China and their relationship with meteorological factors is of great scientific and practical significance. Such analyses provide a foundation for improved monitoring, early warning, and management of high-risk areas and for protecting regional air quality and ecological security.

Satellite remote sensing has increasingly replaced traditional ground-based approaches as a core tool for forest fire monitoring, owing to its wide spatial coverage, high temporal resolution, and rapid response capabilities [11]. Among these instruments, the Moderate Resolution Imaging Spectroradiometer (MODIS) is one of the earliest and most widely used systems for active fire detection. Mounted on the Terra (1999) and Aqua (2002) satellites, MODIS provides 36 spectral bands ranging from visible to thermal infrared, enabling up to four global observations per day. Since 2000, it has generated consistent global fire products at 1 km resolution, creating a valuable long-term record. However, its performance can be limited by cloud cover, smoke, and infrequent overpasses, and it is less effective for detecting small or topographically complex fires. Despite these limitations, MODIS has played a fundamental role in global forest fire monitoring [12,13]. The Visible Infrared Imaging Radiometer Suite (VIIRS), carried on Suomi NPP (2011), NOAA-20 (2017), and NOAA-21 (2022), provides higher spatial resolution (375 m) and improved sensitivity to small-scale and nighttime fires. Although VIIRS offers better spatial precision than MODIS, it remains limited by similar temporal constraints [14]. The Himawari-8/9 geostationary satellites, launched by the Japan Meteorological Agency in 2014 and 2016, are equipped with the Advanced Himawari Imager (AHI). Since 2015, they have provided 10 min interval fire data covering East Asia and the western Pacific, offering unique advantages for capturing rapidly evolving fire events, though their 2 km spatial resolution is relatively coarse. Previous studies have confirmed their accuracy and wide coverage in fire monitoring across Southeast Asia and Australia [15–18].

While polar-orbiting satellites such as MODIS and VIIRS offer high spatial resolution, their limited overpass frequency restricts the detection of diurnal variations and short-lived fires. In contrast, geostationary satellites like Himawari-8/9 provide continuous high-frequency observations that effectively capture fire dynamics but lack fine spatial detail. Integrating both observation types allows the combination of complementary strengths—high spatial accuracy from polar-orbiting sensors and high temporal resolution from geostationary ones. Building on this rationale, the present study focuses on southwestern China by fusing fire hotspot data from Himawari-8/9, MODIS, and VIIRS to construct a high-resolution spatiotemporal fire dataset. Forest hotspots were filtered using land-use information, and Fire Radiative Power (FRP) was integrated to estimate Fire Radiative Energy (FRE). Using the emission factor method, we quantified emissions of ten pollutants (including CO_2, CO, $PM_{2.5}$, VOCs, and NO_X), thereby developing a top-down, multi-pollutant emission inventory for regional forest fires. The Geodetector model was then applied to quantitatively evaluate the explanatory power of meteorological variables and their interactions on spatial emission patterns. Together, these analyses enhance the accuracy of fire monitoring and emission estimation in regions with complex terrain, providing a solid scientific basis for wildfire early warning, carbon management, and air quality protection in southwestern China.

2. Materials and Methods

This study targeted southwestern China (Sichuan, Yunnan, Guizhou, and Chongqing) during the period 2016–2023 (Figure 1). Fire hotspot data were obtained from the NASA Earthdata system for MODIS and VIIRS (https://urs.earthdata.nasa.gov/, accessed on 24 June 2025) and from the Japan Meteorological Agency's P-Tree platform for Himawari-8/9 (https://www.eorc.jaxa.jp/ptree/, accessed on 24 June 2025). The fire products derived from these satellites include geographic coordinates, observation time, fire area, and FRP (MW). FRP, defined as the rate of radiative energy released during combustion, is a key physical parameter for estimating fire emissions from remote sensing. Temporal integration of FRP yields FRE (MJ), which represents the total energy released during the entire burning process. FRE is widely recognized as a robust indicator of fire intensity. Previous research has shown a strong linear relationship between FRE and emissions of multiple atmospheric pollutants, confirming its suitability as a proxy for estimating fire emissions [19,20].

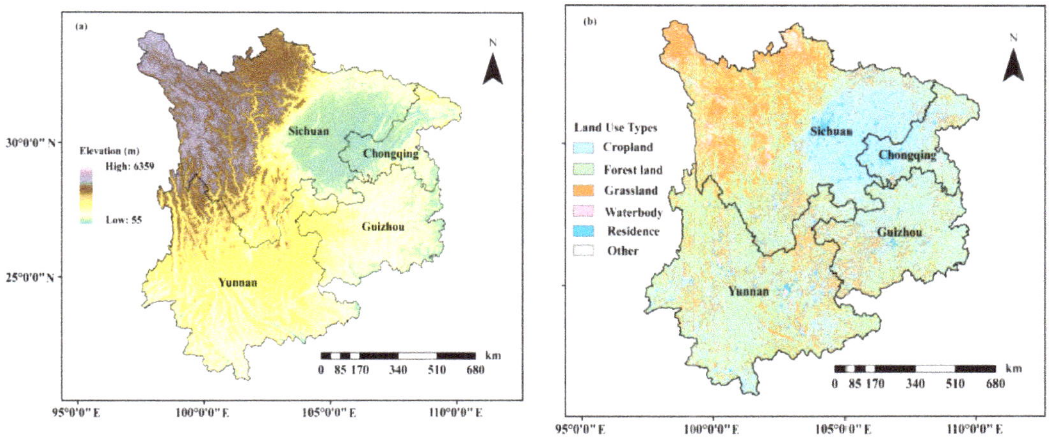

Figure 1. (**a**) Elevation distribution across the four provinces of southwestern China; (**b**) land use type distribution across the four provinces of southwestern China.

Although polar-orbiting satellites such as MODIS and VIIRS provide relatively high spatial resolutions (1 km and 375 m, respectively), their limited daily overpasses hinder continuous monitoring of fire activity and the capture of full diurnal fire cycles. To minimize temporal sampling bias, the diurnal variation in FRP was reconstructed using a Gaussian parameterization proposed by Vermote et al. (2009) [21], which has been widely applied in global fire emission studies to represent sub-daily FRP evolution from MODIS and VIIRS data. The specific formulations used in this study are provided in Equations (1)–(5):

$$\mathrm{FRE} = \int \mathrm{FRP} = \int_0^{24} \mathrm{FRP_{peak}}\left(b + e^{\frac{(t-h)^2}{2\delta^2}}\right)dt \tag{1}$$

$$\mathrm{FRP_{peak}} = \frac{\mathrm{FRP_{Aqura\ day}}}{\left[b + e^{\frac{(13.5\ h)^2}{2\delta^2}}\right]} \tag{2}$$

$$b = 0.86x^2 - 0.52x + 0.08 \quad (3)$$

$$\delta = 3.89x + 1.03 \quad (4)$$

$$h = -\,1.23x + 14.57 + \epsilon \quad (5)$$

In these equations, FRP_{peak} represents the peak fire radiative power of the diurnal cycle; b denotes the background value of radiative energy; σ is the standard deviation of the Gaussian curve; t represents time; and h corresponds to the time of peak FRP. $FRP_{Aqura\ day}$ refers to the mean daytime FRP observed by the Aqua satellite, while x is the ratio of Terra to Aqua monthly mean FRP (T/A). Different vegetation types exhibit distinct T/A values, reflecting their combustion characteristics. For China, this correction factor was set to 4 to adjust the FRP peak time, and the VIIRS diurnal cycle was derived by referencing that of MODIS.

In contrast, the Himawari-8/9 geostationary satellite provides observations every 10 min, offering superior temporal resolution. For this dataset, cumulative FRE was calculated by assuming a uniform FRP distribution between consecutive time steps, as shown in Equation (6):

$$FRE = \int FRP = \int_0^{24} FRPdt \quad (6)$$

Given the high temporal frequency of Himawari-8/9, this approach effectively captures short-term fire fluctuations and eliminates the need to reconstruct a diurnal cycle, as required for polar-orbiting sensors.

For hotspot allocation at a 10 min resolution, a grid-based fusion approach was applied. Following the relative spatial resolutions of the satellite products, a practical priority principle of "VIIRS FRE > MODIS FRE > Himawari-8/9 FRE" was adopted for fire energy fusion at the 10 min grid level [22,23]. This approach reflects the complementary strengths of different sensors: VIIRS provides high spatial resolution, MODIS offers moderate resolution with frequent overpasses, and Himawari-8/9 provides high-temporal-resolution observations. The hotspot with the highest spatial resolution within each grid was selected as the fused hotspot, and the fused FRE for each grid cell was calculated using Equation (7):

$$FRE_{fused} = w_1 \times FRE_H + w_2 \times FRE_M + w_3 \times FRE_V \quad (7)$$

In this equation, FRE_{fused} denotes the fused fire radiative energy on a grid basis, while FRE_H, FRE_M, and FRE_v represent the FRE values calculated from Himawari-8/9, MODIS, and VIIRS FRP data, respectively. The weights w_1, w_2,and w_3 were assigned as follows: when Himawari-8/9 FRE > 0 but MODIS and VIIRS FRE = 0, then $w_1 = 1$, $w_2 = 0$, $w_3 = 0$. when MODIS FRE > 0 and VIIRS FRE = 0, then $w_1 = 0$, $w_2 = 1$, $w_3 = 0$; and when VIIRS FRE > 0,then $w_1 = 0$, $w_2 = 0$, $w_3 = 1$. This weighting strategy is consistent with the sensor fusion logic used in previous multi-satellite emission studies [24], which emphasize exploiting the complementary characteristics of polar-orbiting and geostationary sensors while prioritizing spatial resolution in the fused product.

This fusion strategy prioritizes spatial resolution to ensure higher positional accuracy of detected hotspots, while recognizing that temporal discrepancies among sensors (e.g., between Himawari-8/9's 10 min scans and the less frequent MODIS/VIIRS observations) may influence the representation of short-lived fires. Total pollutant emissions from forest fires, denoted as EI (t), were calculated as follows [25]:

$$EI_{i,\,j} = FRE_{i,\,j} \times EF_{i,\,j}/10^6 \quad (8)$$

where EI represents the emissions of pollutant i in region j (unit: t), EF is the emission factor (unit: g/MJ), and i refers to pollutant types, including CO_2, CO, VOCs, NH_3, SO_2, NOx, $PM_{2.5}$, PM_{10}, OC, and BC.

To filter hotspots, land-use data for southwestern China were obtained from the Resource and Environmental Science Data Center of the Chinese Academy of Sciences (https://www.resdc.cn/, accessed on 27 June 2025). Each hotspot was assigned a land-use category, and only those within forested areas were retained as valid forest fire detections. To ensure temporal consistency, a "nearest-year matching" approach was applied: 2016 was matched with 2015 data, 2017–2018 with 2018 data, 2019–2021 with 2020 data, and 2022–2023 with 2023 data. Distinguishing between flaming and smoldering combustion phases is not feasible with current satellite observations; therefore, unified (composite) emission factors were used to represent overall fire emissions. Current satellite fire products primarily detect thermal anomalies through brightness temperature or FRP, which reflect fire intensity rather than combustion phase [26–28]. Moreover, most forest fires involve canopy and surface fuels that burn with high oxygen availability, making flaming combustion the dominant mode, while smoldering typically occurs only in localized environments such as peat or duff layers [29–31].

Due to the limited availability of emission factor data for forest fires in southwestern China, representative values were derived through a literature synthesis following established biomass-burning inventories [32–35]. Preference was given to studies with vegetation types and combustion conditions similar to those in the study region. Emission factors from multiple sources were averaged to obtain regionally representative values for major fire pollutants (Table 1). This literature-based approach is consistent with global inventories such as GFED and FINN, where FRE is used to estimate total biomass consumption through empirically calibrated coefficients [27,36]. Compared with statistical inventories based solely on burned area, the FRP–FRE approach provides improved temporal resolution and compatibility across multiple sensors, making it suitable for long-term regional emission estimation [28,37].

Table 1. Emission factor of air pollutants from forest fire combustion (g/MJ).

CO_2	CO	$PM_{2.5}$	SO_2	NO_X	OC	VOC_S	NH_3	PM_{10}	BC	Reference
564.20	42.64	/	0.29	1.23	4.47	5.49	0.40	nan	0.23	[32]
543.29	43.87	nan	0.41	1.23	3.75	11.39	0.57	nan	0.23	[33]
543.29	43.87	5.33	0.41	1.23	2.51	8.93	0.57	5.13	0.23	[34]
598.35	35.10	4.81	0.23	1.16	2.85	4.16	0.43	5.31	0.32	Average [37]
552.45	44.01	4.43	0.41	0.68	3.58	7.24	0.89	nan	0.19	Average [35]
560.32	41.90	4.86	0.35	1.11	3.43	7.44	0.57	5.22	0.24	Overall Average

In this study, following references [27,28,36], the conversion coefficient between biomass and energy was adopted as 0.41 kg/MJ.

Meteorological data used in this study were obtained from the ERA5-Land monthly reanalysis dataset published by the European Centre for Medium-Range Weather Forecasts (ECMWF). The data have a spatial resolution of $0.1^\circ \times 0.1^\circ$and include monthly variables such as 2 m air temperature, total precipitation, dew point temperature, and soil moisture. Relative humidity was derived from ERA5 air temperature and dew point data. To ensure spatiotemporal consistency between meteorological and fire hotspot datasets, ERA5 data from the same month and year as fire observations were extracted for each hotspot. Each hotspot's geographic coordinates were mapped to the corresponding ERA5 grid cell to extract the associated meteorological variables. These data were used both for visualization of fire–climate relationships and as input variables for the Geodetector model.

The Geodetector method was employed to quantitatively assess the relationship between forest fire FRE and key meteorological variables, including 2 m air temperature, relative humidity, precipitation, and soil moisture. Geodetector is a statistical tool designed to detect and quantify spatial heterogeneity and its underlying drivers. It evaluates the explanatory power of an independent variable (x) on a dependent variable (y) by comparing the similarity of their spatial distributions. Unlike conventional multivariate methods, Geodetector does not assume linearity or normality, is robust against multicollinearity, and can explicitly capture nonlinear effects and interaction mechanisms. Given the pronounced spatial heterogeneity and complex coupling between fire emissions and meteorological conditions in this study, the Geodetector framework is particularly suitable for accurately identifying and quantifying the explanatory power of both individual and interacting factors. The method has been widely applied and validated in environmental science, geography, and climate research [38,39]. Using Geodetector, we systematically assessed the influence of meteorological factors on the spatiotemporal variability of forest fires in southwestern China.

The analysis was conducted in R (version 4.5.0) using the GD package (version 10.6), applying three supervised discretization methods: equal interval, natural breaks, and quantile. The model automatically selected the optimal stratification scheme within a range of 3–9 classes to maximize the Q-statistic. The model quantifies the explanatory power of meteorological factors on forest fire activity through the Q-statistic, which ranges from 0 to 1 (Equation (9)), with higher values indicating stronger explanatory power. The Q-statistic can be applied to stratification across space, time, or attributes, enabling effective identification of driving factors at different scales [38].

$$Q = 1 - \frac{\sum_{h=1}^{L} N_h \sigma_h^2}{N\sigma^2} \tag{9}$$

In the equation, h = 1...L denotes the stratification of the dependent variable (e.g., fire radiative energy, FRE, and the number of fire hotspots) or the independent variable (e.g., meteorological factors). N_h and N represent the sample size in stratum h and the total sample size in the entire study area (or fire-affected area), respectively. ${\sigma_h}^2$ and σ^2 denote the variance of the dependent variable within stratum h and across the entire region, respectively.

3. Results

3.1. Comparison Analysis

To evaluate the performance and complementarity of multi-satellite collaborative monitoring, this study conducted a comprehensive comparative analysis of the performance of Himawari-8/9, MODIS, and VIIRS in detecting forest fire hotspots across southwestern China, focusing on both interannual dynamics and spatial patterns (Figure 2). In terms of interannual variability, Himawari-8/9, with its 10 min observation frequency, detected an average of 405,800 hotspots annually from 2016 to 2023, comprising 95.41% of the total. Its high temporal resolution made it particularly effective during high-risk fire years such as 2022, when it substantially outperformed VIIRS (16,000 hotspots annually, 3.76%) and MODIS (3500 hotspots annually, 0.83%). Leveraging its 375 m spatial resolution, VIIRS excelled at identifying small and scattered surface fires in mountainous terrain, while MODIS utilized its multispectral information to effectively distinguish combustion phases and surface temperatures. Although the three sensors exhibited consistent interannual variation trends, differences in spatial resolution, detection thresholds, and sensitivity led to a complementary monitoring framework, wherein each sensor provided unique advantages despite quantitative discrepancies.

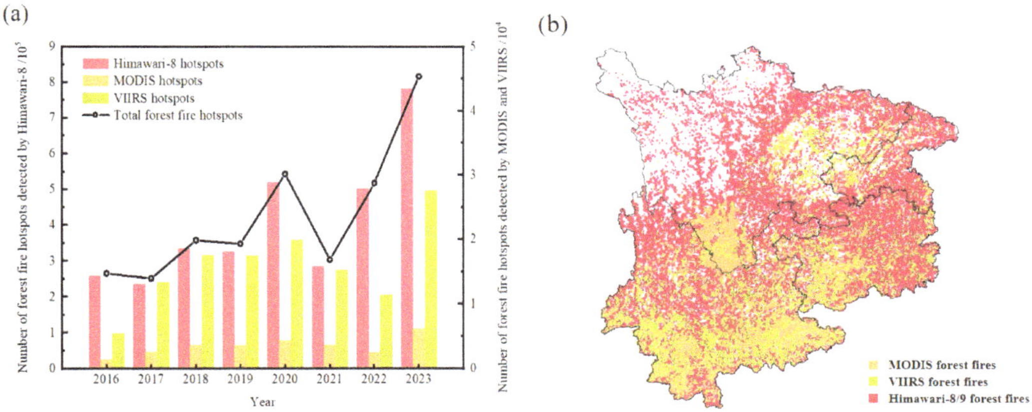

Figure 2. Interannual variation in forest fire hotspots (**a**) and spatial distribution of forest fire hotspots (**b**) monitored by Himawari-8/9, MODIS, and VIIRS in southwestern China during 2016–2023.

Regarding spatial performance, despite its coarse 2 km spatial resolution, Himawari 8/9's high-frequency imaging ensured continuous coverage over complex terrain and improved detection of both persistent and short-lived intense fires. This temporal advantage reduced omission errors and enhanced emission estimation accuracy by capturing short-duration events that might otherwise be missed. Compared to VIIRS (with a 12 h revisit time) and MODIS (with four daily overpasses), Himawari-8 significantly increased hotspot detection in cloud-prone mountainous areas, thus mitigating the underestimation of total emissions common in single-satellite approaches. While VIIRS and MODIS improved local detection accuracy and resilience to cloud shadow interference, their limited temporal resolution restricted their ability to capture complete fire dynamics. Overall, Himawari-8/9's high-frequency observations greatly expanded the spatiotemporal coverage of fire hotspot detection, providing critical support for the construction of continuous and comprehensive fire emission inventories. This finding underscores the importance of multi-satellite synergy in wildfire monitoring and carbon emission estimation in topographically complex regions.

3.2. Emission Inventory of Air Pollutants

The emission inventory developed through multi-source satellite hotspot fusion (Table 2) reveals that forest fires in southwestern China emitted ten major pollutants from 2016 to 2023: CO_2, CO, $PM_{2.5}$, PM_{10}, VOCs, NOx, SO_2, NH_3, BC, and OC. Their average annual emissions and associated standard deviations ($\times 10^3$ t·a^{-1}) were 5623.58 ± 1554.33, 356.84 ± 98.63, 41.39 ± 11.44, 44.46 ± 12.29, 63.36 ± 17.51, 9.45 ± 2.61, 2.98 ± 0.82, 4.86 ± 1.34, 2.04 ± 0.56, and 29.21 ± 8.08, respectively. Among these, CO_2 and CO dominated total gaseous emissions, whereas particulate matter ($PM_{2.5}$, PM_{10}) and carbonaceous components (BC, OC), though smaller in quantity, exerted disproportionately large impacts on visibility, human health, and radiative forcing. Reactive gases such as NOx, SO_2, NH_3, and VOCs also played crucial roles as precursors of ozone and secondary aerosols.

Table 2. Pollutant emissions from forest fires in southwestern China during 2016–2023 ($\times 10^3$ t).

Year		CO	NOx	SO_2	NH_3	VOCs	$PM_{2.5}$	PM_{10}	BC	OC	CO_2
2016	Guizhou	29.29	0.78	0.24	0.40	5.20	3.40	3.65	0.17	2.40	461.5
	Sichuan	59.01	1.56	0.49	0.80	10.48	6.84	7.35	0.34	4.83	929.9
	Yunnan	233.70	6.19	1.95	3.18	41.50	27.11	29.11	1.34	19.13	3682.9
	Chongqing	4.25	0.11	0.04	0.06	0.75	0.49	0.53	0.02	0.35	66.98
	Total	326.25	8.64	2.73	4.44	57.93	37.84	40.64	1.87	26.71	5141.5
2017	Guizhou	32.69	0.87	0.27	0.44	5.80	3.79	4.07	0.19	2.68	515.1
	Sichuan	63.03	1.67	0.53	0.86	11.19	7.31	7.85	0.36	5.16	993.3
	Yunnan	161.58	4.28	1.35	2.20	28.69	18.74	20.13	0.93	13.23	2546.4
	Chongqing	4.05	0.11	0.03	0.06	0.72	0.47	0.50	0.02	0.33	63.83
	Total	261.35	6.92	2.18	3.56	46.41	30.31	32.56	1.50	21.39	4118.7
2018	Guizhou	64.56	1.71	0.54	0.88	11.46	7.49	8.04	0.37	5.28	1017.4
	Sichuan	69.52	1.84	0.58	0.95	12.34	8.06	8.66	0.40	5.69	1095.6
	Yunnan	159.69	4.23	1.33	2.17	28.36	18.52	19.89	0.91	13.07	2516.6
	Chongqing	3.21	0.09	0.03	0.04	0.57	0.37	0.40	0.02	0.26	50.59
	Total	296.98	7.87	2.48	4.04	52.73	34.45	37.00	1.70	24.31	4680.2
2019	Guizhou	45.44	1.20	0.38	0.62	8.07	5.27	5.66	0.26	3.72	716.1
	Sichuan	55.83	1.48	0.47	0.76	9.91	6.48	6.96	0.32	4.57	879.8
	Yunnan	240.31	6.37	2.01	3.27	42.67	27.87	29.94	1.38	19.67	3787.1
	Chongqing	2.62	0.07	0.02	0.04	0.47	0.30	0.33	0.02	0.21	41.29
	Total	344.20	9.12	2.88	4.68	61.12	39.92	42.88	1.97	28.18	5424.3
2020	Guizhou	76.25	2.02	0.64	1.04	13.54	8.84	9.50	0.44	6.24	1201.6
	Sichuan	118.61	3.14	0.99	1.61	21.06	13.76	14.78	0.68	9.71	1869.1
	Yunnan	219.03	5.80	1.83	2.98	38.89	25.41	27.29	1.25	17.93	3451.8
	Chongqing	1.97	0.05	0.02	0.03	0.35	0.23	0.25	0.01	0.16	31.05
	Total	415.86	11.02	3.47	5.66	73.84	48.24	51.81	2.38	34.04	6553.6
2021	Guizhou	93.83	2.49	0.78	1.28	16.66	10.88	11.69	0.54	7.68	1478.7
	Sichuan	95.84	2.54	0.80	1.30	17.02	11.12	11.94	0.55	7.85	1510.3
	Yunnan	151.55	4.01	1.27	2.06	26.91	17.58	18.88	0.87	12.41	2388.3
	Chongqing	2.03	0.05	0.02	0.03	0.36	0.24	0.25	0.01	0.17	31.99
	Total	343.25	9.09	2.87	4.67	60.95	39.81	42.76	1.97	28.10	5409.4
2022	Guizhou	59.60	1.58	0.50	0.81	10.58	6.91	7.43	0.34	4.88	939.2
	Sichuan	72.47	1.92	0.61	0.99	12.87	8.41	9.03	0.42	5.93	1142.0
	Yunnan	152.61	4.04	1.27	2.08	27.10	17.70	19.01	0.87	12.49	2405.0
	Chongqing	9.27	0.25	0.08	0.13	1.65	1.08	1.15	0.05	0.76	146.0
	Total	293.95	7.79	2.46	4.00	52.20	34.10	36.62	1.68	24.06	4632.4
2023	Guizhou	78.61	2.08	0.66	1.07	13.96	9.12	9.79	0.45	6.44	1238.8
	Sichuan	108.94	2.89	0.91	1.48	19.34	12.64	13.57	0.62	8.92	1716.8
	Yunnan	383.00	10.15	3.20	5.21	68.01	44.42	47.72	2.19	31.35	6035.8
	Chongqing	2.33	0.06	0.02	0.03	0.41	0.27	0.29	0.01	0.19	36.72
	Total	572.88	15.18	4.79	7.79	101.72	66.45	71.37	3.28	46.90	9028.2

The interannual emissions of pollutants exhibited a generally fluctuating upward trend, with 2020 and 2023 identified as years of intense fire activity when most pollutants peaked. For example, CO emissions in Sichuan Province rose markedly from 55.83 $\times$ 10^3 t in 2019 to 118.61 $\times$ 10^3 t in 2020, an increase of 62.78 $\times$ 10^3 t (approximately 112%). As a result, the total regional CO emissions rose from 344.20 $\times$ 10^3 t in 2019 to 415.86 $\times$ 10^3 t in 2020, an increase of 71.66 $\times$ 10^3 t. The provincial shares of CO emissions that year were: Yunnan 52.7%, Sichuan 28.5%, Guizhou 18.3%, and Chongqing 0.5%. In Sichuan, multiple pollutants (including $PM_{2.5}$, PM_{10}, and CO_2) increased concurrently, consistent with the occurrence of extreme fire events. A representative case was the "3·30" Xichang wildfire in Liangshan Prefecture, which burned over 1000 ha and caused 19 fatalities.

In 2023, regional CO emissions rose sharply again, rising from 293.95 × 10^3 t in 2022 to 572.88 × 10^3 t, an increase of 278.93 × 10^3 t. Yunnan Province contributed most to this surge, with its CO emissions surging from 152.61 × 10^3 t to 383.00 × 10^3 t (an increase of 230.39 × 10^3 t), accounting for 66.9% of the regional total in 2023. This escalation coincided with several large-scale wildfires in Yunnan that year, such as the "4·11" fire in Jiangchuan, Yuxi (mobilizing 3000–4000 personnel for firefighting, with open flames extinguished between April 15–16), and the April 13 Anning wildfire.

A provincial comparison revealed that pollutant emissions from forest fires in southwestern China followed the pattern "Yunnan > Sichuan > Guizhou > Chongqing." Yunnan accounted for more than half of the total annual emissions due to its dense forests, rugged terrain, and pronounced dry season. Sichuan exhibited a "low-frequency but high-intensity" pattern, with elevated emissions during extreme events. Guizhou displayed moderate levels, and Chongqing consistently recorded the lowest emissions, reflecting its limited forest coverage and urbanized landscape.

3.3. Comparison of Hotspot Numbers and Emission Characteristics

As a typical emission from forest fires, CO_2 has a strong cumulative effect in the atmosphere and serves as an important indicator for evaluating the environmental impacts of fires. To further characterize the emission patterns and regional differences in forest fires in southwestern China, we analyzed the spatial distribution of fire hotspots and total CO_2 emissions, as well as their proportions in Yunnan, Sichuan, Guizhou, and Chongqing from 2016 to 2023 (Figure 3). Results indicated that, during the study period, the annual average number of forest fire hotspots in the region was 4.25 × 10^5, while the annual average CO_2 emissions amounted to 5.62 × 10^{12} g (approximately 5.62 × 10^6 t), showing an overall upward but fluctuating trend.

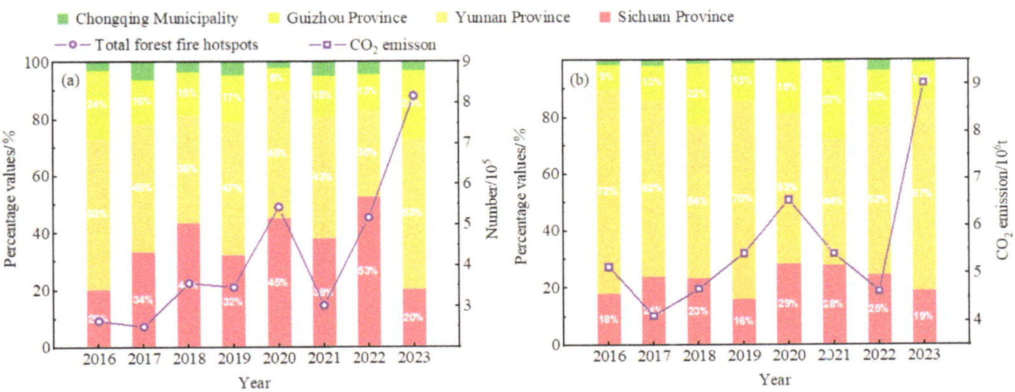

Figure 3. Proportion and total amounts of forest fire hotspots (**a**) and proportion and total amounts of CO_2 emissions (**b**) in southwestern China from 2016 to 2023.

In terms of provincial distribution, the proportions of fire hotspots were 47.8% in Yunnan, 30.4% in Sichuan, 17.6% in Guizhou, and 4.2% in Chongqing. Corresponding proportions of CO_2 emissions were 59.6%, 22.5%, 16.8%, and 1.0%, respectively. Yunnan consistently ranked first in both hotspot count and emissions across most years, confirming its role as the dominant contributor to regional carbon output. This dominance is closely associated with its dense forest coverage, complex topography, and seasonally dry climate.

Sichuan Province exhibited pronounced interannual variation in fire emissions. In 2020, influenced by the catastrophic "3·30" Xichang wildfire, its CO_2 share rose sharply to 28.5%, far exceeding normal levels and underscoring the significant role of extreme fire events in driving regional carbon emissions. In contrast, Guizhou maintained relatively stable proportions of hotspots and emissions, averaging 17.6% and 13.5%, respectively, indicating a moderate contribution to overall emissions. Chongqing, constrained by limited forest resources and extensive urbanization, consistently recorded the lowest values in both indicators, reflecting its minor impact on regional wildfire emissions.

Further comparison revealed that Yunnan's share of CO_2 emissions exceeded its hotspot proportion by approximately 11.4 percentage points, implying higher per-hotspot carbon intensity and stronger combustion efficiency. In contrast, Guizhou and Chongqing exhibited lower CO_2 emission shares relative to hotspot proportions, suggesting smaller fire scales and weaker combustion strength. In 2023, the number of forest fire hotspots in the region increased from 517,773 in 2022 to 816,585, a rise of 57.71%. During the same period, CO_2 emissions almost doubled, increasing from 4.63×10^{12} g to 9.03×10^{12} g, an increase of 95.0%. The emission growth far outpaced the increase in hotspot numbers, suggesting that not only did fire frequency rise, but combustion intensity and FRP also increased substantially, leading to higher per-hotspot carbon emissions. These findings highlight that satellite-based evaluations of forest fire emissions should incorporate multiple parameters—such as FRP, fire duration, and burned area—rather than relying solely on hotspot counts to capture true environmental impacts.

3.4. Characteristics Spatiotemporal Characteristics of Carbon Emissions

At the prefecture-level scale (Figure 4), forest fires in Sichuan Province were mainly concentrated in Panzhihua and Liangshan Prefecture; in Yunnan Province, high-frequency fire areas included Lijiang, Xishuangbanna, Pu'er, Wenshan, and Dehong; in Guizhou Province, fires frequently occurred in Qianxinan, Qiannan, Qiandongnan Prefectures, and Anshun City; while Chongqing Municipality generally exhibited relatively low fire frequencies.

From a geographical perspective, the Sichuan Basin, characterized by flat terrain and extensive croplands, recorded relatively few fire hotspots. In contrast, fire activity was concentrated around the basin's periphery, the Yunnan–Guizhou Plateau, southern Guizhou, and the Hengduan Mountains. These areas, featuring high forest coverage, rugged terrain, and complex climatic conditions, provided favorable conditions for frequent fire occurrence. Overall, the spatial distribution of hotspots followed the pattern of "more in the south than in the north, and more in mountainous areas than in plains." Mountainous regions exhibited greater sensitivity to fire occurrence, identifying them as priority zones for regional fire prevention and control.

Figure 5 illustrates the multi-scale variations in CO_2 emissions from forest fires in southwestern China between 2016 and 2023. At the interannual scale, Yunnan Province peaked in 2023, with emissions significantly exceeding previous years, likely linked to exceptional drought and heat conditions. Sichuan Province experienced a sharp increase in 2020, primarily due to the "3·30" Xichang wildfire, underscoring the substantial effect of extreme events on annual carbon fluxes. In contrast, Guizhou and Chongqing maintained low emission levels with relatively stable year-to-year changes.

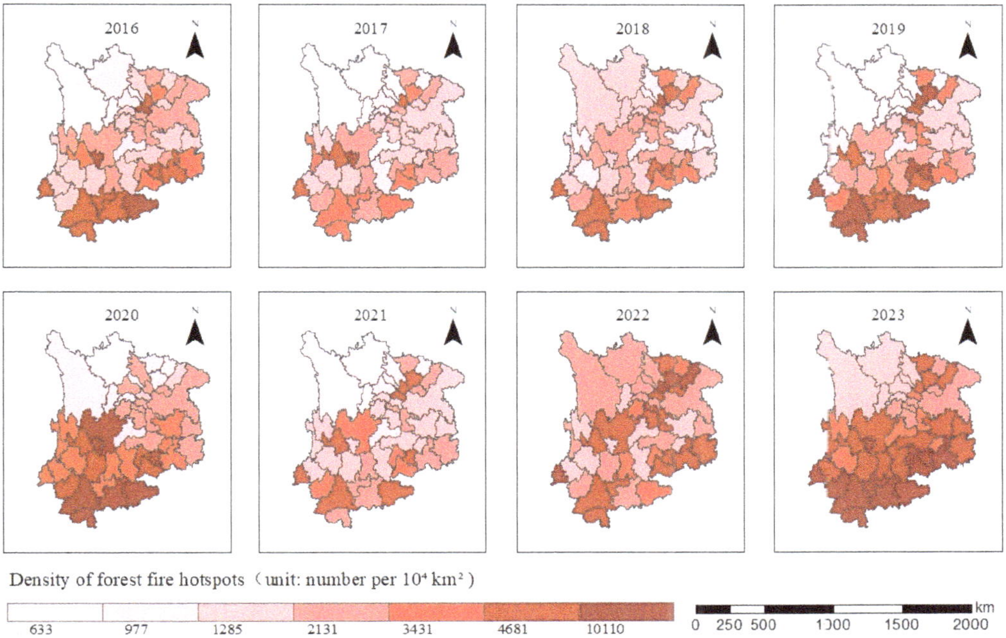

Figure 4. Distribution of forest fire hotspot density across prefecture-level cities in southwestern China from 2016 to 2023.

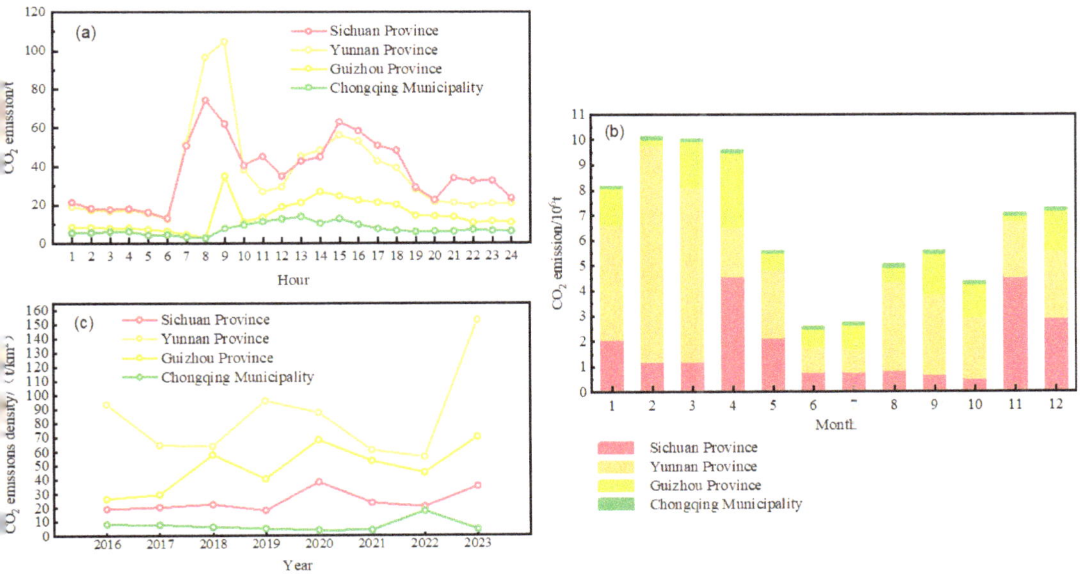

Figure 5. Diurnal (**a**), monthly (**b**), and annual (**c**) variations in CO_2 emissions from forest fires in southwest China during 2016–2023.

At the intra-annual scale, CO_2 emissions exhibited marked seasonal concentration. Emissions were mainly concentrated in January–April and November–December, but the peak months differed by province—February in Yunnan, April in Sichuan, and March in Guizhou—reflecting the strong influence of spring drought conditions on fire activity. During summer (June–September), enhanced precipitation and humidity suppressed emissions, narrowing regional differences.

At the diurnal scale, CO_2 emissions followed a distinct pattern of "low at night, high during the day, and decreasing in the evening." Between 0:00 and 6:00, emissions remained generally low, with most provinces reaching their minimum around 6:00. From 08:00 onward, emissions increased rapidly as the temperature rose and the humidity decreased, peaking at around 09:00 in Yunnan and Guizhou and around 15:00 in Sichuan. The high-emission phase generally persisted until 19:00, after which emissions gradually decreased with cooling temperatures.

Overall, the results demonstrate that forest fire emissions in southwestern China exhibited pronounced interannual fluctuations, seasonal concentration, and diurnal variability, driven jointly by climatic conditions and human activities. Provincial differences were evident: Yunnan peaked mainly in February, Sichuan in April, and Guizhou in March. At the diurnal scale, Yunnan and Guizhou peaked near 09:00, while Sichuan peaked later at 15:00. It is therefore recommended that high-emission provinces such as Yunnan and Sichuan establish province-specific, season-specific, and time-targeted monitoring and prevention systems tailored to their emission characteristics.

3.5. Meteorological Drivers and Their Impacts on Carbon Emissions

Temperature and precipitation were selected as key meteorological variables to analyze their relationships with forest fire CO_2 emissions at both interannual and intra-annual (monthly) scales. February is the peak fire season in southwestern China. Figure 6a illustrates the interannual variations in temperature, precipitation, and CO_2 emissions during February from 2016 to 2023. At the interannual scale, CO_2 emissions generally followed the pattern of "higher temperatures resulting in higher emissions, whereas increased precipitation corresponded to lower emissions," confirming the enhancing role of hot and dry conditions on fire intensity and carbon emissions. However, this relationship was nonlinear. For instance, during 2019–2020, despite rising temperatures, abundant precipitation corresponded to relatively low CO_2 emissions, implying that complex nonlinear and interactive effects among meteorological variables jointly influenced fire development and emission strength. At the intra-annual (monthly) scale (Figure 6b), from January to March, low precipitation and dry atmospheric conditions led to a sharp increase in CO_2 emissions with rising temperatures, reflecting the intensified combustion driven by hot dry conditions. From April to July, although temperatures continued to climb, enhanced rainfall suppressed emissions, indicating that moist conditions could partially offset the fire-promoting influence of heat. From August to December, as temperatures declined and precipitation decreased again, CO_2 emissions rebounded, showing that dryness re-emerged as the dominant driver.

FRP, representing the instantaneous rate of energy release from combustion, correlates strongly with fire intensity and biomass consumption rate. Compared with estimates based solely on burned area or emission inventories, satellite-derived FRP provides real-time data with high spatial and temporal resolution and clear physical comparability, offering a more objective measure of fire strength and spatial variability. Accordingly, this study employed temperature (T), precipitation (P), relative humidity (RH), and soil moisture (SM) as independent variables and applied the Geodetector model to assess their explanatory power on the spatial distribution of FRP (Figure 7). In the Geodetector framework, Q (0–1) represents the proportion of FRP variance explained by a single factor or by a factor pair, with interpretation based on statistical significance and relative magnitude rather than fixed thresholds [40]. The overall ranking indicated a dominance of moisture-related factors, with mean contributions during 2016–2023 of RH (0.1223) > P (0.1103) ≈ SM (0.1083) > T (0.0750). This result suggests that indicators describing the atmospheric and fuel moisture state (RH, SM) were stronger determinants of fire intensity, while temperature alone exerted a comparatively weaker influence. All variables passed the significance test ($p < 0.05$).

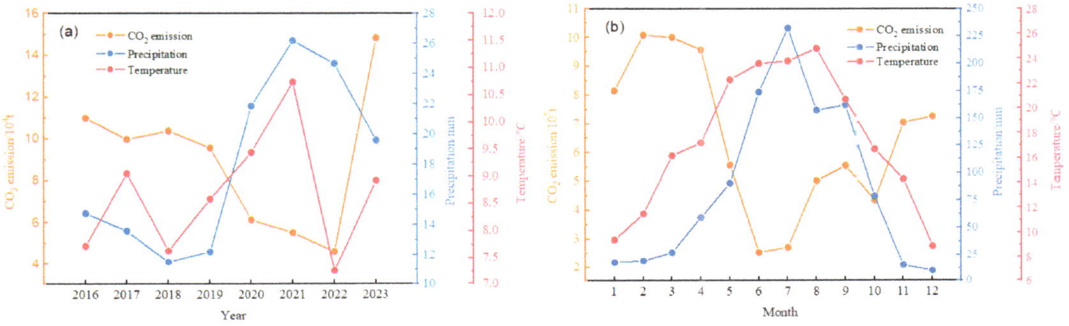

Figure 6. Interannual (**a**) and monthly (**b**) variations in CO_2 emissions from forest fires in southwestern China, in relation to temperature and precipitation.

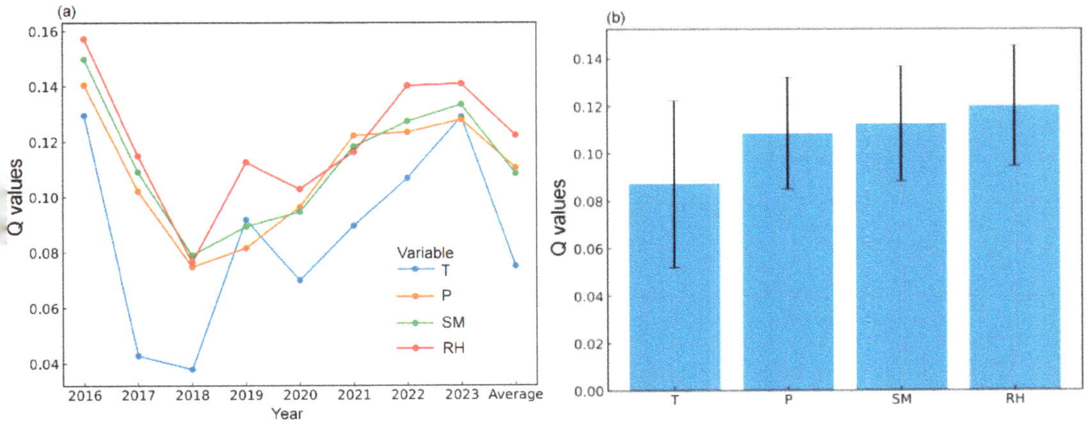

Figure 7. Yearly and average factor detector Q values (**a**) and mean ± SD of factor detector Q values (**b**) of meteorological factors (2016–2023).

At the interannual scale, a fluctuating pattern of "high in 2016—low in 2018—rebound in 2023" was observed. The mean Q value of the four factors peaked at 0.144 in 2016, declined to 0.067 in 2018, remained low during 2019–2020, and then gradually increased to 0.133 in 2023. For individual factors, all four exhibited troughs in 2018 (T = 0.038, P = 0.074, SM = 0.079, RH = 0.076), while 2016 showed widespread peaks (T = 0.130, P = 0.140, SM = 0.150, RH = 0.157), indicating that hydrothermal conditions that year exerted the strongest differentiation effect on the spatial heterogeneity of FRP. Among the factors, temperature (T) showed the largest interannual fluctuation ($\sigma = 0.035$), significantly higher than precipitation (P, 0.023), soil moisture (SM, 0.024), and relative humidity (RH, 0.025). In terms of linear trends, the Q values of all four factors increased slowly, suggesting that after the 2018 low point, their explanatory power gradually recovered and strengthened. Overall, moisture-related factors (RH, SM, and P) provided stable contributions, whereas temperature was more sensitive to extreme years, amplifying or weakening its explanatory power for FRP spatial distribution.

Interaction detection results revealed widespread significant bi-factor enhancement effects among the four meteorological variables, all of which elevated the explanatory power (Q values) for FRP spatial distribution in southwestern China beyond single-factor levels (Figure 8). Among the meteorological interactions, T∩RH was the most prominent, with a multi-year average Q = 0.1486 (σ = 0.0604), exceeding both RH (0.1223, σ = 0.022) and T (0.0750, σ = 0.028), and exhibiting the greatest interannual variability. This indicates that the synergistic amplification effect of "high temperature–low humidity" varied substantially across years. For instance, in 2017, T∩RH reached 0.2983, about 2.6 times higher than the 2019 low of 0.1128, demonstrating its dominant role in shaping FRP spatial heterogeneity under anomalously hot dry conditions. By contrast, in 2019, overall climatic forcing was weak, and interaction effects were generally low.

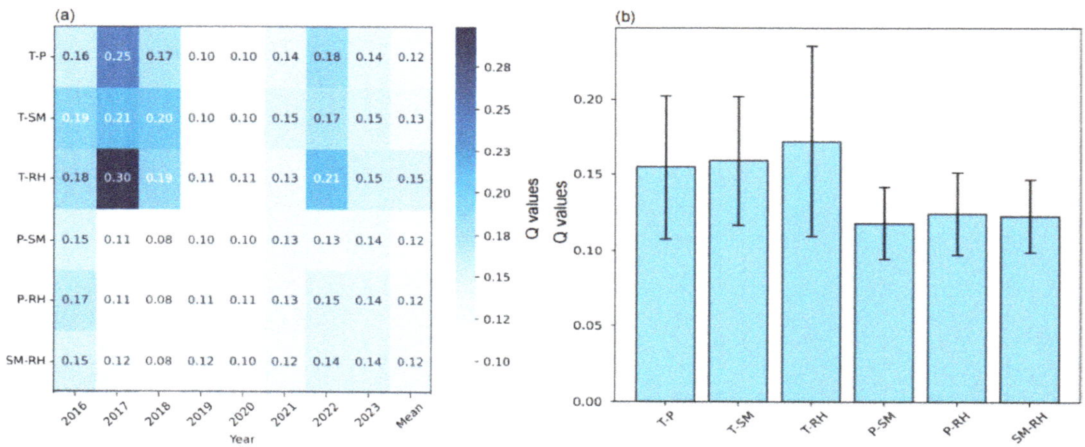

Figure 8. Yearly and average interaction detector Q values (**a**) and mean ± SD of interaction detector Q values (**b**) of meteorological factor pairs (2016–2023).

In the Geodetector framework, the relationship between Q(T∩RH) and the individual factors (Q(T), Q(RH)) reveals the interaction type. When Q(T∩RH) > max[Q(T), Q(RH)], it represents a bi-factor enhancement, meaning the two variables jointly increase explanatory power beyond their independent effects. When Q(T∩RH) > Q(T) + Q(RH), it reflects a non-linear enhancement, where the combined influence exceeds additive effects, indicating strong coupling and feedback. In this study, T∩RH = 0.1486 exceeded both T (0.0750) and RH (0.1223) and passed the significance test ($p < 0.05$), showing a clear bi-factor enhancement between temperature and humidity. In 2017, Q(T∩RH) even surpassed Q(T) + Q(RH), indicating non-linear enhancement under extreme hot dry conditions.

Physically, high temperature enhances evapotranspiration and heating, while low humidity directly reduces fuel moisture content; their coupling forms a hydrothermal synergy that intensifies fuel desiccation and FRP variability. This mechanism aligns with established findings that hot and dry conditions (high vapor pressure deficit, VPD) jointly lower live fuel moisture and promote extreme fire behavior [41–43].

Beyond T∩RH, both T∩SM (mean 0.1310, σ = 0.037) and T∩P (0.1218, σ = 0.047) also showed enhancement effects with pronounced fluctuations: higher in 2017 and 2022 (e.g., 2022 values of T∩RH = 0.2115, T∩SM = 0.1742, P∩RH = 0.1466), but lower during 2018–2019 (e.g., only T∩P was relatively high in 2018 at 0.1726, while other combinations declined; in 2019, T∩P = 0.0994, T∩SM = 0.1002). By comparison, P∩RH (0.1246, σ = 0.027) and SM∩RH (0.1233, σ = 0.027) were more stable and moderately enhanced, reflecting a background modulation role.

Overall, T∩RH emerged as the dominant interaction (highest mean and largest variability), while T∩SM and T∩P served as secondary but sensitive factors, with their explanatory power significantly enhanced under low water availability or anomalous years (e.g., 2017, 2022). In contrast, P∩RH and SM∩RH remained relatively stable. Mechanistically, RH, SM, and P directly reflect the wet–dry status of fuels and environment, determining fuel moisture content and combustibility. Temperature indirectly reduces fuel moisture by intensifying evapotranspiration and heating, making its influence more dependent on coupling with moisture conditions, consistent with the strong effect of T∩RH. From a management and early warning perspective, fire risk assessment and energy monitoring should avoid reliance on single factors, and instead adopt modeling frameworks that capture interactions and interannual non-stationarity. This provides a quantitative basis for fire risk management and intensity regulation under coupled hydrothermal conditions in southwestern China.

Looking ahead, climate projections indicate that warming and aridification will further heighten forest-fire risks in southwestern China. Model simulations suggest that rising temperatures and decreasing humidity will extend both the duration and intensity of high-fire-danger periods. Fan et al. [44] showed that escalating hot dry extremes amplify compound fire-weather risk globally. At the national scale, Shao et al. [45] predicted significantly higher fire occurrence probabilities across China under future climate scenarios, with temperature and precipitation identified as strong climatic controls.

These projections reinforce our findings that temperature and relative humidity are the key meteorological controls, and their coupling (hot dry synergy) exerts the strongest enhancement on fire radiative power. The expected intensification of compound heat–drought events will likely increase both fire frequency and CO_2 emission magnitude, posing greater challenges for regional greenhouse gas mitigation and China's carbon-peaking goals.

4. Discussion

4.1. Comparison with Database-Based Emission Results

The Global Fire Emissions Database, version 5 (GFED5), jointly developed by several international research institutions, is a global fire emission inventory constructed using a "bottom-up" approach. It employs MODIS burned area as the core input, corrected with higher-resolution data from Sentinel-2 and Landsat, and integrates vegetation cover, fuel load, combustion completeness, and emission factors to estimate fire emissions. GFED5 provides global emission data at 0.25° and monthly resolution, covering a wide range of gases and aerosol components, and serves as a key dataset for climate change studies, carbon cycle assessments, and atmospheric chemistry modeling [39]. For comparison, emission estimates corresponding to boreal and temperate forest categories in GFED5 were selected and clipped using the geographical extent of southwestern China to extract regional forest fire emissions.

The comparison of forest fire emissions in southwestern China during 2016–2022 (Figure 9) shows that, although both this study and GFED5 capture broadly consistent interannual variations, noticeable discrepancies remain in emission magnitudes. For example, in 2017, this study reported substantially higher estimates of CO, NH_3, $PM_{2.5}$, BC, and OC than GFED5, marking a departure from the general consistency observed across years. Overall, our estimates of CO, NH_3, $PM_{2.5}$, BC, and OC exceeded those of GFED5, whereas GFED5 generally yielded higher values of NOx and CO_2, with only minor differences for SO_2. These divergences mainly arise from the choice of emission factors and satellite observation strategies. In this study, region-specific emission factors were employed to reflect combustion conditions in southwestern China; for instance, the CO factor (102.2 g/kg) and OC factor (8.37 g/kg) were higher than the global averages used in GFED5 (90.5 g/kg and 7.155 g/kg, respectively), better capturing incomplete combustion characteristics and thereby elevating CO, OC, and $PM_{2.5}$ estimates [46]. Furthermore, the integration of Himawari-8/9, MODIS, and VIIRS data enables improved detection of small-scale and short-duration fires, enhancing the representation of low-temperature combustion products such as NH_3 and OC. This multi-satellite fusion approach offers a more complete characterization of regional fire activity, particularly in the complex terrain of southwestern China.

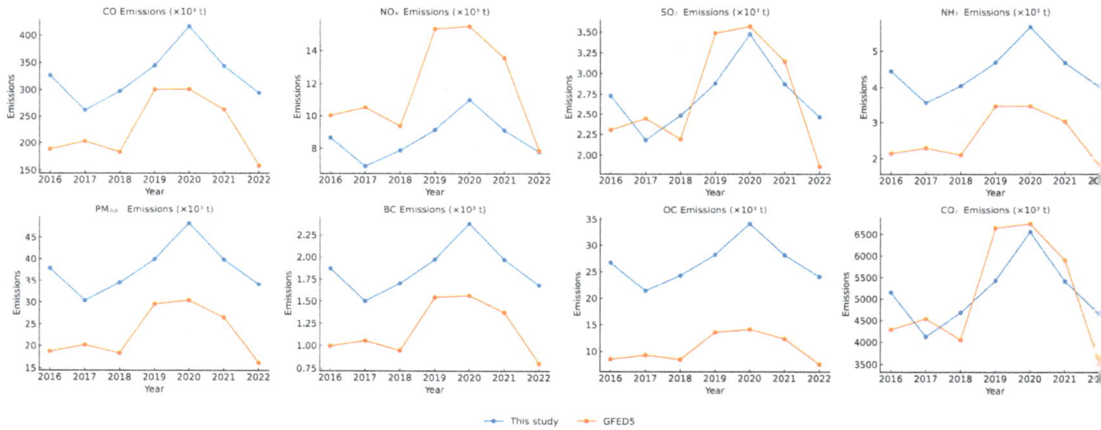

Figure 9. Comparison of pollutant emissions from forest fires in southwest China with GFED5.

By contrast, GFED5's higher estimates of NOx and CO_2 likely result from its globally uniform emission factors and stronger capability to capture high-intensity crown fires, under the assumption of more complete combustion. This assumption may lead to overestimation of CO_2 emissions in open valleys or during dry years. For instance, the NOx emission factor used in GFED5 is based on global averages and does not sufficiently account for the mixed combustion characteristics typical of southwestern China, leading to potential overestimation of NOx emissions under high-temperature, complete combustion conditions. Overall, GFED5 emphasizes global consistency and model universality, while this study focuses more on regional adaptability, observational completeness, and improved small-scale fire detection. These differences underscore the value of our inventory for regional applications and highlight considerations for integrating region-specific emission factors and higher-resolution input data into global models.

Direct validation with ground or airborne measurements remains limited due to scarce field data in southwestern China. To provide a relative evaluation, our estimates were compared with GFED5, showing consistent temporal trends but differing emission magnitudes due to varying satellite inputs and emission factors.

Preliminary tests with hourly meteorological data revealed that short-term fluctuations introduced substantial noise and reduced model stability. Therefore, monthly ERA5-Land data were adopted, balancing computational feasibility and the ability to capture dominant climatic controls on fire activity. While this temporal resolution may smooth short-lived extremes such as heatwaves or wind bursts, it remains suitable for regional and long-term analyses. Future work should incorporate higher-temporal-resolution meteorological datasets and multi-source observations to improve cross-validation and reduce uncertainties in emission estimation.

4.2. Uncertainty Analysis

This study estimated forest fire pollutant emissions in southwestern China based on multi-source satellite observations from Himawari-8/9, MODIS, and VIIRS. Although synergistic observations enhanced the spatiotemporal coverage of fire detection, heterogeneity among datasets and model assumptions introduced substantial uncertainties. The three satellites differ markedly in observation frequency and spatial resolution. The adopted priority rule ("VIIRS > MODIS > Himawari") ensures higher spatial precision while filling temporal gaps using Himawari-8/9's high-frequency observations. However, this approach may still bias the representation of short-lived fires. Specifically, when only Himawari-8/9 detects transient or subpixel fires that are undetected by MODIS or VIIRS, their radiative energy and burned area may be underestimated due to the coarser spatial resolution of Himawari-8/9. Conversely, finer-resolution detections from MODIS/VIIRS may lead to slight FRE overestimation if temporal alignment with Himawari-8/9 is imperfect. Such temporal–spatial mismatches have been reported to cause FRE biases in regions with frequent small-scale fires [14,24]. In addition, cloud cover and topographic shadow effects influence each sensor differently, further amplifying spatial heterogeneity and omission or commission errors in fire detection.

The emission factors used in this study were primarily derived from literature and regional characteristics, but no differentiation was made among fire types detected by different satellites. For example, VIIRS frequently captures small-scale surface fires at 375 m resolution, which may have lower carbon emission efficiency than large-scale crown fires observed by Himawari-8/9. A uniform set of emission factors cannot fully capture these differences, resulting in biases between estimated emissions and actual combustion intensity. In addition, discrepancies in FRP retrieval algorithms among the satellites may introduce systematic biases for the same fire pixels, affecting the consistency of FRE calculations. Recent studies have further demonstrated that uncertainties in biomass burning emissions can arise from variations in fire detection products, combustion phases, and fuel conditions. Hua et al. [47] highlighted that estimates of CO and OC emissions differ markedly among inventories due to inconsistencies in fuel loading and burning conditions. Minsavage-Davis and Davies [48] showed that the accuracy of fire behavior models depends strongly on fuel characteristics and fire spread dynamics, implying that combustion efficiency may vary substantially across vegetation types. Similarly, Zheng et al. [49] found that seasonal shifts from flaming- to smoldering-dominated burning alter the temporal profiles of CO emissions, suggesting that fixed emission factors may not adequately represent such dynamic fire behavior.

In FRE computation, the Himawari-8/9 data were processed under the assumption of a uniform distribution of FRP between consecutive observations, while MODIS and VIIRS data employed Gaussian-based diurnal cycle parameterizations. These simplifications may not fully capture rapid fluctuations in FRP or short-lived fire events, potentially leading to uncertainties in cumulative FRE. For example, transient peaks in combustion intensity within the 10 min Himawari-8/9 sampling interval are approximated as linear changes, while the coarse revisit frequency of MODIS and VIIRS may smooth or miss sub-hourly fire peaks. Nevertheless, Gaussian diurnal reconstruction has been widely applied in global fire emission inventories to compensate for the sparse temporal coverage of polar-orbiting satellites, while linear interpolation between high-frequency geostationary observations remains a common approach for continuous energy estimation. Intercomparisons between polar-orbiting and geostationary satellites have demonstrated that these parameterizations yield consistent FRE estimates despite differences in temporal resolution [50–52]. Therefore, while we acknowledge that such assumptions may smooth short-term fire dynamics, they follow established and widely accepted practices. Future work will integrate higher-temporal-resolution datasets to better represent transient fire behavior and improve FRE accuracy.

5. Conclusions

(a) From 2016 to 2023, the annual average emissions of ten major pollutants from forest fires in southwestern China were CO_2 (5623.58 ± 1554.33), CO (356.84 ± 98.63), $PM_{2.5}$ (41.39 ± 11.44), PM_{10} (44.46 ± 12.29), VOCs (63.36 ± 17.51), NOx (9.45 ± 2.61), SO_2 (2.98 ± 0.82), NH_3 (4.86 ± 1.34), BC (2.04 ± 0.56), and OC (29.21 ± 8.08) (units: $\times 10^3$ $t{\cdot}a^{-1}$). Among these, CO_2 and CO were the dominant pollutants.

(b) Spatially, Yunnan Province and Sichuan Province were the core contributors to CO_2 emissions. Strengthening monitoring, early warning, and emission control in these two provinces is critical for maintaining regional carbon balance and improving air quality. Temporally, CO_2 emissions were higher during daytime than at night and were concentrated between January–April and December.

(c) CO_2 emissions increased with rising temperature and decreased with precipitation, exhibiting the overall pattern of "high temperature promotes emissions, while humidity suppresses fires." The emissions demonstrated nonlinear and interactive relationships with meteorological factors, with the strongest interaction between temperature and relative humidity in winter. Under cold-dry conditions, elevated temperatures combined with low humidity were more likely to trigger fires and amplify carbon emissions.

Author Contributions: L.F.: Methodology, validation, writing original draft preparation. Y.H.: software, investigation. J.L.: Formal analysis, data gathering, review, and editing. W.G.: Conceptualization, supervision, validation, review, and editing. All authors have read and agreed to the published version of the manuscript.

Funding: This research was supported by "the Fundamental Research Funds for the Central Universities" [No. 2682024CX102].

Institutional Review Board Statement: Not applicable.

Informed Consent Statement: Not applicable.

Data Availability Statement: Data is contained within the article.

Conflicts of Interest: The authors declare no conflicts of interest.

Abbreviations

The following abbreviations are used in this manuscript:

AHI	Advanced Himawari Imager
BC	Black Carbon
CO_2	Carbon Dioxide
ECMWF	European Centre for Medium-Range Weather Forecasts
ERA5-Land	ECMWF Re-Analysis 5-Land
FRP	Fire Radiative Power
FRE	Fire Radiative Energy
GFED5	Global Fire Emissions Database version 5
MODIS	Moderate Resolution Imaging Spectroradiometer
NH_3	Ammonia
NO_X	Nitrogen Oxides ($NO + NO_2$)
OC	Organic Carbon
$PM_{2.5}$	Fine Particulate Matter (aerodynamic diameter ≤ 2.5 μm)
PM_{10}	Inhalable Particulate Matter (aerodynamic diameter ≤ 10 μm)
RH	Relative Humidity
SO_2	Sulfur Dioxide
VOCs	Volatile Organic Compounds
VIIRS	Visible Infrared Imaging Radiometer Suite

ferences

IPCC. Climate Change 2021: The Physical Science Basis. In *Contribution of Working Group I to the Sixth Assessment Report of the Intergovernmental Panel on Climate Change*; Cambridge University Press: Cambridge, UK, 2021.

Pimm, S.L.; Raven, P. Extinction by numbers. *Nature* **2000**, *403*, 843–845. [CrossRef]

Garcia-Hurtado, E.; Pey, J.; Borrás, E.; Sánchez, P.; Vera, T.; Carratalá, A.; Alastuey, A.; Querol, X.; Vallejo, V.R. Atmospheric PM and volatile organic compounds released from Mediterranean shrubland wildfires. *Atmos. Environ.* **2014**, *89*, 303–311. [CrossRef]

Guo, Y.X.; Wei, M.; Tian, M.Y.; Sun, X.Y.; Zheng, W.X.; Ma, Y.F.; Guo, F.T. Characteristics of smoke and particulate matter released by combustion of eight main tree species in Yunnan Province. *Res. Environ. Sci.* **2021**, *34*, 2295–2305.

Park, C.Y.; Takahashi, K.; Fujimori, S.; Jansakoo, T.; Burton, C.; Huang, H.; Kou-Giesbrecht, S.; Reyer, C.P.O.; Mengel, M.; Burke, E.; et al. Attributing human mortality from fire $PM_{2.5}$ to climate change. *Nat. Clim. Change* **2024**, *14*, 1193–1200. [CrossRef]

Wang, M.; Jiang, C.; Sun, O.J. Spatially differentiated changes in regional climate and underlying drivers in southwestern China. *J. For. Res.* **2022**, *33*, 755–765. [CrossRef]

Tian, X.R.; Zhao, F.J.; Shu, L.F.; Wang, M.Y. Hotspots from satellite monitoring and forest fire weather index analysis for Southwest China. *For. Res.* **2010**, *23*, 523–529.

Ning, X.C.; Li, J.W.; Zhuang, P.K.; Lai, S.F.; Zheng, X.G. Wildfire combustion emission inventory in Southwest China (2001–2020) based on MODIS fire radiative energy data. *Atmos. Pollut. Res.* **2024**, *15*, 102279. [CrossRef]

He, J.; Zhang, Y. Analysis of the impacts of meteorological factors on forest fire danger in the conterminous United States. *J. Geophys. Res. Atmos.* **2021**, *126*, e2021JD035376. [CrossRef]

Lund, M.T.; Nordling, K.; Gjelsvik, A.B.; Samset, B.H. The influence of variability on fire weather conditions in high latitude regions under present and future global warming. *Environ. Res. Commun.* **2023**, *5*, 065016. [CrossRef]

Feng, Y.; Lei, C.F.; Pan, Y.; Zhang, C.J. Forest fire monitoring based on satellite remote sensing technology. *Geomat. Technol. Equip.* **2022**, *24*, 15–19.

Justice, C.O.; Vermote, E.; Townshend, J.R.G.; Defries, R.; Roy, D.P.; Hall, D.K. The Moderate Resolution Imaging Spectroradiometer (MODIS): Land remote sensing for global change research. *IEEE Trans. Geosci. Remote Sens.* **1998**, *36*, 1228–1249. [CrossRef]

Giglio, L.; Descloitres, J.; Justice, C.O.; Kaufman, Y.J. An enhanced contextual fire detection algorithm for MODIS. *Remote Sens. Environ.* **2003**, *87*, 273–282. [CrossRef]

Schroeder, W.; Oliva, P.; Giglio, L.; Csiszar, I.A. The new VIIRS 375 m active fire detection data product: Algorithm description and initial assessment. *Remote Sens. Environ.* **2014**, *143*, 85–96. [CrossRef]

Chen, J.; Lv, Q.; Wu, S.; Zeng, Y.; Li, M.; Chen, Z.; Zhou, E.; Zheng, W.; Liu, C.; Chen, X.; et al. An adapted hourly Himawari-8 fire product for China: Principle, methodology and verification. *Earth Syst. Sci. Data* **2023**, *15*, 1911–1931. [CrossRef]

16. Lu, X.M.; Zhang, X.Y.; Li, F.J.; Cochrane, M.A. Improved estimation of fire particulate emissions using a combination of VII and AHI data for Indonesia during 2015–2020. *Remote Sens. Environ.* **2022**, *281*, 113238. [CrossRef]
17. Zheng, Y.M.; Liu, J.; Jian, H.D.; Fan, X.T.; Yan, F.L. Fire diurnal cycle derived from a combination of the Himawari-8 and VII satellites to improve fire emission assessments in Southeast Australia. *Remote Sens.* **2021**, *13*, 2852. [CrossRef]
18. Wickramasinghe, C.H.; Wallace, L.; Reinke, K.; Jones, S. Intercomparison of Himawari-8 AHI-FSA with MODIS and VIIRS act fire products. *Int. J. Digit. Earth* **2020**, *13*, 457–473. [CrossRef]
19. Wooster, M.J.; Roberts, G.; Perry, G.L.W.; Kaufman, Y.J. Retrieval of biomass combustion rates and totals from fire radiat power observations: FRP derivation and calibration relationships between biomass consumption and fire radiative energy rele *J. Geophys. Res. Atmos.* **2005**, *110*, D24311. [CrossRef]
20. Liu, M.X.; Song, Y.; Yao, H.; Kang, Y.N.; Li, M.M.; Huang, X.; Hu, M. Estimating emissions from agricultural fires in the No China Plain based on MODIS fire radiative power. *Atmos. Environ.* **2015**, *112*, 326–334. [CrossRef]
21. Vermote, E.; Ellicott, E.; Dubovik, O.; Lapyonok, T.; Chin, M.; Giglio, L.; Roberts, G.J. An approach to estimate global biom burning emissions of organic and black carbon from MODIS fire radiative power. *J. Geophys. Res. Atmos.* **2009**, *114*, D182 [CrossRef]
22. Xu, Y.Q.; Huang, Z.J.; Ou, J.M.; Jia, G.L.; Wu, L.L.; Liu, H.L.; Lu, M.H.; Fan, M.; Wei, J.; Chen, L.F.; et al. Near-real-time estimat of hourly open biomass burning emissions in China using multiple satellite retrievals. *Sci. Total Environ.* **2022**, *817*, 1527 [CrossRef]
23. Li, R.; Liu, J.X.; Zhang, M.D.; Huang, C.W.; Zhao, S.H.; Zhang, M.G.; Cheng, L.F. Development and analysis of a forest emission inventory in China considering daily variation patterns. *Environ. Pollut.* **2025**, *382*, 126720. [CrossRef] [PubMed]
24. Kang, Y.; Im, J. Mitigating underestimation of fire emissions from the Advanced Himawari Imager: A machine learning a multi-satellite ensemble approach. *Int. J. Appl. Earth Obs. Geoinf.* **2024**, *128*, 103784. [CrossRef]
25. Xu, Y.Q. Study on the Evolution Characteristics and Dynamic Representation of Open Biomass Burning Emissions in China. Ph Thesis, South China University of Technology, Guangzhou, China, 2020.
26. Anderson, L.D.; Dix, B.; Schnell, J.; Yokelson, R.; Veefkind, J.P.; Ahmadov, R.; de Gouw, J. Analyzing the impact of evolv combustion conditions on the composition of wildfire emissions using satellite data. *Geophys. Res. Lett.* **2023**, *50*, e2023GL105 [CrossRef]
27. Roy, D.P.; Boschetti, L.; Smith, A.M.S. Satellite remote sensing of fires. In *Fire Phenomena and the Earth System*; Wiley: Hoboken, USA, 2013; Chapter 5. [CrossRef]
28. Santoso, M.A. Review of the transition from smouldering to flaming combustion in wildfires. *Front. Mech. Eng.* **2019**, *5*, [CrossRef]
29. Garg, P.; Wang, S.; Oakes, J.M.; Bellini, C.; Gollner, M.J. Variations in gaseous and particulate emissions from flaming a smouldering combustion of Douglas fir and lodgepole pine under different fuel moisture conditions. *Combust. Flame* **20** *263*, 113386. [CrossRef]
30. Keeley, J.E. Fire intensity, fire severity and burn severity: A brief review and suggested usage. *Int. J. Wildland Fire* **2009**, 116–126. [CrossRef]
31. Fernandes, P.M.; Botelho, H.S. A review of prescribed burning effectiveness in fire hazard reduction. *Int. J. Wildland Fire* **2003**, 117–128. [CrossRef]
32. Andreae, M.O. Emission of trace gases and aerosols from biomass burning—An updated assessment. *Atmos. Chem. Phys.* **20** *19*, 8523–8546. [CrossRef]
33. Streets, D.G.; Bond, T.C.; Carmichael, G.R.; Fernandes, S.D.; Fu, Q.; He, D.; Klimont, Z.; Nelson, S.M.; Tsai, N.Y.; Wang, M.Q.; et An inventory of gaseous and primary aerosol emissions in Asia in the year 2000. *J. Geophys. Res. Atmos.* **2003**, *108*, D21. [CrossR
34. Andreae, M.O.; Merlet, P. Emission of trace gases and aerosols from biomass burning. *Glob. Biogeochem. Cycles* **2001**, *15*, 955–9 [CrossRef]
35. Wu, J.; Kong, S.F.; Wu, F.Q.; Cheng, Y.; Zheng, S.R.; Yan, Q.; Zheng, H.; Yang, G.W.; Zheng, M.M.; Liu, D.T.; et al. Estimati the open biomass burning emissions in central and eastern China from 2003 to 2015. *Atmos. Chem. Phys.* **2018**, *18*, 11623–116 [CrossRef]
36. Freeborn, P.H.; Wooster, M.J.; Hao, W.M.; Ryan, C.A.; Nordgren, B.L.; Baker, S.P.; Ichoku, C. Relationships between energy relea fuel mass loss, and trace gas and aerosol emissions during laboratory biomass fires. *J. Geophys. Res. Atmos.* **2008**, *113*, D013 [CrossRef]
37. Song, Y.; Liu, B.; Miao, W.J.; Chang, D.; Zhang, Y.H. Spatiotemporal variation in nonagricultural open fire emissions in Chi from 2000 to 2007. *Glob. Biogeochem. Cycles* **2009**, *23*, GB2008. [CrossRef]
38. Fan, H.; Yang, X.; Zhao, C.; Yang, Y.K.; Shen, Z.Y. Spatiotemporal variation characteristics of global fires and their emissio *Atmos. Chem. Phys.* **2023**, *23*, 7781–7798. [CrossRef]

Chen, Y.; Hall, J.; van Wees, D.; Andela, N.; Hantson, S.; Giglio, L.; van der Werf, G.R.; Morton, D.C.; Randerson, J.T. Multi-decadal trends and variability in burned area from the fifth version of the Global Fire Emissions Database (GFED5). *Earth Syst. Sci. Data* **2023**, *15*, 5227–5259. [CrossRef]
Wang, J.F.; Xu, C.D. Geodetector: Principle and prospective. *Acta Geogr. Sin.* **2017**, *72*, 116–134. [CrossRef]
Abatzoglou, J.T.; Williams, A.P. Impact of anthropogenic climate change on wildfire across western US forests. *Proc. Natl. Acad. Sci. USA* **2016**, *113*, 11770–11775. [CrossRef]
Pimont, F.; Ruffault, J.; Martin-StPaul, N.K.; Dupuy, J.-L. Why is the effect of live fuel moisture content on fire rate of spread underestimated in field experiments in shrublands? *Int. J. Wildland Fire* **2019**, *28*, 127–137. [CrossRef]
Ma, W.; Zhai, L.; Pivovaroff, A.; Shuman, J.; Buotte, P.; Ding, J.; Christoffersen, B.; Knox, R.; Moritz, M.; Fisher, R.A.; et al. Assessing climate change impacts on live fuel moisture and wildfire risk using a hydrodynamic vegetation model. *Biogeosci. Discuss.* **2021**, *18*, 4005–4020. [CrossRef]
Fan, X.W.; Miao, C.Y.; Zscheischler, J.; Slater, L.; Wu, Y.; Chai, Y.; Aghakouchak, A. Escalating Hot-Dry Extremes Amplify Compound Fire Weather Risk. *Earth's Future* **2023**, *11*, e2023EF003976. [CrossRef]
Shao, Y.K.; Fan, G.P.; Feng, Z.K.; Sun, L.H.; Yang, X.H.; Ma, T.T.; Li, X.S.; Fu, H.N.; Wang, A.A. Prediction of Forest Fire Occurrence in China under Climate Change Scenarios. *J. For. Res.* **2023**, *34*, 1217–1228. [CrossRef]
Van der Werf, G.R.; Randerson, J.T.; Giglio, L.; Van Leeuwen, T.T.; Chen, Y.; Rogers, B.M.; Mu, M.; Van Marle, M.J.E.; Morton, D.C.; Collatz, G.J.; et al. Global fire emissions estimates during 1997–2016. *Earth Syst. Sci. Data* **2017**, *9*, 697–720. [CrossRef]
Hua, W.X.; Lou, S.J.; Huang, X.; Xue, L.; Ding, K.; Wang, Z.; Ding, A.J. Diagnosing uncertainties in global biomass burning emission inventories and their impact on modeled air pollutants. *Atmos. Chem. Phys.* **2024**, *24*, 6787–6807. [CrossRef]
Minsavage-Davis, C.D.; Davies, G.M. Evaluating the Performance of Fire Rate of Spread Models in Northern-European Calluna vulgaris Heathlands. *Fire* **2022**, *5*, 46. [CrossRef]
Zheng, B.; Chevallier, F.; Yin, Y.; Wang, Y.L. On the role of the flaming-to-smoldering transition in the seasonal cycle of African fire emissions. *Geophys. Res. Lett.* **2018**, *45*, 6633–6641. [CrossRef]
Kaiser, J.W.; Heil, A.; Andreae, M.O.; Benedetti, A.; Chubarova, N.; Jones, L.; Morcrette, J.-J.; Razinger, M.; Schultz, M.G.; Suttie, M.; et al. Biomass burning emissions estimated with a global fire assimilation system based on observed fire radiative power. *Biogeosciences* **2012**, *9*, 527–554. [CrossRef]
Andela, N.; Morton, D.C.; Giglio, L.; Chen, Y.; Van der Werf, G.R.; Kasibhatla, P.S.; DeFries, R.S.; Collatz, G.J.; Hantson, S.; Kloster, S.; et al. A human-driven decline in global burned area. *Science* **2017**, *356*, 1356–1362. [CrossRef]
Roberts, G.; Wooster, M.J.; Lagoudakis, E. Annual and diurnal African biomass burning temporal dynamics. *Biogeosciences* **2009**, *6*, 849–866. [CrossRef]

environments

MDPI

Article

Can Saharan Dust Intrusions Alter the Isotopic Composition of Atmospheric Methane and Carbon Dioxide?

Isabella Zaccardo [1,2,*], Antonella Buono [1,2,*], Teresa Laurita [1], Benedetto De Rosa [1], Francesco D'Amico [3,4], Davide Amodio [1], Francesco Cardellicchio [1], Canio Colangelo [1], Gianluca Di Fiore [1], Giuseppe Gargano [1], Aldo Giunta [1], Emilio Lapenna [1], Ermann Ripepi [1], Marco Rosoldi [1], Donato Summa [1] Serena Trippetta [1], Claudia Roberta Calidonna [3], Alcide Giorgio di Sarra [5], Guido Masiello [2] and Lucia Mona [1]

1 Institute of Methodologies for Environmental Analysis, National Research Council of Italy (CNR-IMAA), Contrada Santa Loja, Tito Scalo, I-85050 Potenza, Italy; teresa.laurita@cnr.it (T.L.); benebetto.derosa@cnr.it (B.D.R.); davide.amodio@cnr.it (D.A.); francesco.cardellicchio@cnr.it (F.C.); nicolagianluca.difiore@cnr.it (G.D.F.); giuseppe.gargano@cnr.it (G.G.); aldo.giunta@cnr.it (A.G.); emilio.lapenna@cnr.it (E.L.); ermann.ripepi@cnr.it (E.R.); marco.rosoldi@cnr.it (M.R.); donato.summa@cnr.it (D.S.); serena.trippetta@cnr.it (S.T.); lucia.mona@cnr.it (L.M.)

2 Department of Engineering, University of Basilicata (Unibas DiING), Via dell'Ateneo Lucano 10, I-85100 Potenza, Italy; guido.masiello@unibas.it

3 Institute of Atmospheric Sciences and Climate, National Research Council of Italy (CNR-ISAC), Area Industriale Comparto 15, Lamezia Terme, I-88046 Catanzaro, Italy; francescodamico2@cnr.it (F.D.); claudiaroberta.calidonna@cnr.it (C.R.C.)

4 Department of Biology, Ecology and Earth Sciences, University of Calabria (Unical DiBEST), Via Pietro Bucci Cubo 15B, Rende, I-87036 Cosenza, Italy

5 Italian National Agency for New Technologies, Energy and Sustainable Economic Development (ENEA), Via Enrico Fermi 45, Frascati, I-00044 Rome, Italy; alcide.disarra@enea.it

* Correspondence: isabellazaccardo@cnr.it (I.Z.); antonellabuono@cnr.it (A.B.)

Abstract

This study investigates the potential influence of Saharan dust intrusions on the stable carbon isotopes of methane (CH_4) and carbon dioxide (CO_2), using continuous in situ measurements from the ICOS (Integrated Carbon Observation System) atmospheric station at the CNR-IMAA Atmospheric Observatory (CIAO) in Tito, Basilicata, southern Italy. The analysis builds upon the recent investigation of a two-month isotopic dataset (20 February–20 April 2025), which identified anomalous isotopic behavior in March, coinciding with three distinct dust events. The observations reveal shifts in $\delta^{13}C$–CH_4 values that align temporally with the dust intrusions, accompanied by a decrease in CH_4 mole fractions. Such patterns could suggest fractionation processes affecting CH_4, potentially driven by enhanced oxidation promoted by mineral aerosols. At the same time, $\delta^{13}C$–CO_2 shows a gradual decline, deviating from the typical springtime enrichment associated with intensified photosynthetic uptake of $^{12}CO_2$. This unexpected decrease suggests that dust-related radiative effects, particularly the attenuation of incoming solar radiation, may inhibit photosynthesis, thereby altering the isotopic composition of atmospheric CO_2. Consistently, CO_2 mole fractions exhibit a modest increase during periods of declining $\delta^{13}C$–CO_2, reinforcing the interpretation of reduced photosynthetic activity. These findings provide new observational data beyond existing studies from Atlantic regions and highlight the need for further research on the role of mineral dust in shaping greenhouse gas isotopic variability in the Mediterranean, where such events are frequent and climatically relevant.

Keywords: Saharan dust intrusion; stable carbon isotopes; $\delta^{13}C$ fractionation; atmospheric methane

1. Introduction

CH_4 is a potent and photochemically active greenhouse gas (GHG) with an atmospheric lifetime of approximately 11.2 ± 1.3 years [1]. Its global warming potential (GWP) far exceeds that of CO_2 over shorter timeframes, about 80 times greater over 20 years (GWP20) and 28 times greater over 100 years (GWP100) [2]. This makes CH_4 a critical target for near-term climate mitigation strategies. Since 2007, atmospheric CH_4 concentrations have been rising steadily [3–5], reflecting a disequilibrium between its sources and sinks [6]. The increase followed a plateau of several years, during which no major fluctuations in terms of concentrations and isotopic fingerprint were observed [7]. CH_4 is emitted from biogenic, anthropogenic, and geological sources, while its removal is primarily driven by oxidation via hydroxyl radicals (OH•), chlorine atoms (Cl), and methanotrophic bacteria [8–11]. These oxidation processes alter the stable carbon isotope ratio ($\delta^{13}C$) of CH_4, as lighter isotopes ($^{12}CH_4$) are preferentially consumed, enriching the atmosphere in $^{13}CH_4$ [12,13]. Biogenic sources such as wetlands, termites, ruminants, and rice paddies typically emit CH_4 that is depleted in ^{13}C, with $\delta^{13}C$–CH_4 values ranging from approximately −65‰ to −55‰ compared to the international VPDB (Vienna Pee Dee Belemnite) standard. In contrast, fossil fuel combustion and biomass burning produce CH_4 with less negative $\delta^{13}C$–CH_4 values, generally ranging from −45‰ to −20‰ compared to the VPDB. The resulting atmospheric mixture of emissions yields an average $\delta^{13}C$–CH_4 value around −54.3‰. Isotopic analysis of CH_4 thus provides a powerful tool for source apportionment and for constraining the global CH_4 budget [14,15]. However, the $\delta^{13}C$–CH_4 observed in the atmosphere also reflects isotopic fractionation during oxidation. This enrichment effect shifts $\delta^{13}C$–CH_4 toward less negative values, around −47.5‰ [16], potentially leading to an underestimation of biogenic contributions in top-down isotope-constrained models [17]. Moreover, uncertainties in atmospheric Cl concentrations [18] propagate through these models, affecting the accuracy of CH_4 source attribution [10].

Recent studies have highlighted the role of mineral dust aerosols, particularly Saharan dust, in modulating CH_4 oxidation [19]. These aerosols provide reactive surfaces that facilitate chemical transformations capable of altering $\delta^{13}C$-CH_4 signatures. Saharan dust, a major component of atmospheric aerosols, exerts wide-reaching effects on climate, ecosystems, and biogeochemical cycles. Mona et al. [20] provided a comprehensive overview of these impacts, including modifications to the Earth's radiation budget, changes in cloud microphysics, and influences on biogeochemical cycles. Their work also emphasized the rapid evolution of dust observational capabilities, spanning in situ measurements, ground-based networks, and satellite remote sensing. More recently, advances in field instrumentation have enabled deeper investigation into how mineral aerosol transported by Saharan dust influences the oxidative capacity of the atmosphere and impacts trace gas concentrations [21].

Saharan dust is known to carry nutrients such as iron (Fe) and phosphorus (P); when mineral dust travelling across the ocean mixes with sea spray aerosol containing sodium chloride (NaCl), it forms mineral dust sea-spray aerosol (MDSA), acting as a large surface on which chemical reactions can occur [22]. Under sunlight, Fe in mineral dust acts as a catalyst, triggering chemical reactions with chloride ions (Cl^-) to produce Cl•, highly reactive gas-phase Cl radicals capable of modifying the oxidizing capacity of the atmosphere [23–25]. This mechanism may lead to an increase in $\delta^{13}C$-CH_4 due to the kinetic isotope effect during CH_4 oxidation [26–28]. This process develops over a time span ranging from a few hours to several days and may mask the decreasing $\delta^{13}C$-CH_4 signatures from biogenic and fossil fuel sources, complicating the interpretation of the drivers behind the observed increase in atmospheric CH_4. A work by D'Amico et al. [29] highlighted the potential of using

https://doi.org/10.3390/environments13030145

$\delta^{13}C-CH_4$ fractionation to perform local assessments of the Cl sink, because the magnitude of the Cl-initiated CH_4 sink remains poorly constrained, and its overall contribution to regional CH_4 loss is still the subject of active investigation [9,10]. Inspired by natural mechanisms like MDSA, the Iron-Salt Aerosol (ISA) method [30] has been proposed as a geoengineering strategy to enhance atmospheric CH_4 removal. This approach involves lofting Fe-rich aerosols to catalyze Cl radical formation, thereby accelerating CH_4 oxidation. While promising, the ISA method requires further investigation to assess its feasibility, environmental impact, and safety.

A recent study by Buono et al. [31] analyzed isotopic measurements collected over a 60-day period (20 February–20 April) at the newly established ICOS [32] atmospheric station in southern Italy for greenhouse gas observations, named POT station. Their results showed that $\delta^{13}C-CH_4$ and $\delta^{13}C-CO_2$ values in March deviated notably from those in February and April. This anomaly was interpreted as a possible indication of the Saharan dust influence on the isotopic composition of atmospheric CH_4 and CO_2. Although the study did not directly quantify the impact of mineral aerosols, it provided preliminary evidence that episodic dust intrusions may modulate CH_4 isotopic signatures, potentially through enhanced oxidative processes involving Cl radicals. Building on these observations, the present study aims to investigate this phenomenon further by correlating isotopic shifts with documented dust intrusion events, searching to identify potential links between aerosol dynamics and changes in $\delta^{13}C$ signatures. Future research will be necessary to confirm the underlying chemical mechanisms, including the role of reactive halogen species in CH_4 oxidation.

In this work, we analyze data from continuous measurements of stable carbon isotopes in CH_4 and CO_2 performed during three Saharan dust intrusions between 10 and 26 March 2025. Observations were conducted at the POT atmospheric station [33] near Potenza, southern Italy, part of CIAO and currently under evaluation for ICOS labeling. The study integrates isotopic measurements with aerosol data from in situ instruments and ground-based remote sensing techniques [34,35].

These results represent a fundamental step toward expanding high-resolution GHG monitoring in southern Europe. A recent study [36] emphasized the strategic relevance of the Potenza site in addressing spatial gaps in the ICOS network. The isotopic fingerprinting technique pioneered by Keeling [37] is directly employed in this study, its conceptual framework may be essential for interpreting CH_4 and CO_2 variability in the central Mediterranean.

2. Materials and Methods

2.1. GHGs Observation

2.1.1. Sampling System

The POT station, located on a plain surrounded by low mountains in Southern Italy (Tito, 40.60° N, 15.72° E, 760 m a.s.l.), is less than 150 km from the Mediterranean Sea's coasts in the west, south, and east directions. This strategic location places it within the central Mediterranean Basin, approximately 560 km north of the African continent (S-SW direction), and makes it an ideal site for studying Saharan dust [38,39]. The ICOS station is equipped with a 104 m-high tower featuring three air intake levels at 10, 50 and 100 m, with an advanced air sampling system. At the base of the tower, a climate-controlled shelter houses an array of ICOS compliant instruments dedicated to monitoring greenhouse gases.

For isotopic measurements, the air is sampled at 100 m above ground level (a.g.l), to reduce the influence of local sources [40], minimize the impact of soil respiration [41], and ensure that the data are representative of regional atmospheric conditions. The air sampling system begins with an EATON Synflex 1300 (EATON, Dublin, Ireland) tube with an outer

 https://doi.org/10.3390/environments1303

diameter (OD) of 12 mm that extends from the 100 m intake point down to the shelter. The tube connects to a KNF N 815 KTE (KNF, Schenkon, Switzerland) flushing pump to ensure efficient air transport. Inside the shelter, the sampling line transitions to an EATON Synflex 1300 tube with a reduced OD of 6 mm via Swagelok fittings. The air is filtered through a first universal filter with a porosity of 2 µm (M&C F2, Ratingen, Germany) to remove particulates and a second in-line Swagelok filter with a porosity of 0.5 µm (Swagelok, Solon, OH, USA) prior to reaching the analyzer. Before analysis, gathered air is dried using a Perma Pure MD 070-144 S-4 Nafion dryer (Perma Pure, Lakewood, NJ, USA) to ensure moisture-free measurements. Ambient air is sampled with a flow rate of about 25 mL/min.

2.1.2. GHGs Instrumentation

For the purpose of this study, greenhouse gas measurements were performed using a Picarro G2201-i (Santa Clara, CA, USA), operating in simultaneous CO_2–CH_4 mode. This instrument enables continuous monitoring of mole fractions (ppm) of CO_2 and CH_4, along with their respective carbon isotope ratios ($\delta^{13}C$-CO_2 and $\delta^{13}C$-CH_4), referenced to the VPDB standard [42,43]. The analyzer achieves a precision of <0.16‰ for $\delta^{13}C$-CO_2 and <1.15‰ for $\delta^{13}C$-CH_4 in High Precision (HP) mode, which was selected based on the observed CH_4 concentration range (1.8–12 ppm). The instrument performs one measurement per second, allowing high-frequency data acquisition suitable for capturing fine-scale atmospheric variability. CRDS technology, on which its operation is based, ensures reliable and continuous measurements with minimal calibration needs [44]. However, maintaining data quality requires strict control of key instrumental parameters. During operation, cavity pressure is kept within 148 Torr, while cavity and warm box temperatures are maintained within ±0.003 °C of the 45 °C setpoint. These parameters are continuously monitored, and deviations may indicate potential performance issues.

The dataset analyzed in this study is publicly available through the ITINERIS (Italian Integrated Environmental Research Infrastructures System) HUB [45]. Data acquisition and evaluation follow procedures described in Buono et al. [30]. In that work, the performance and stability of the Picarro G2201-i analyzer were assessed by comparing its dry mole fraction measurements with those obtained from a co-located Picarro G2401 (Santa Clara, CA, USA). The comparison, based on 1015 h of overlapping observations, was performed using the Bland & Altman method [46–48], confirming strong agreement between the two instruments and validating the reliability of the G2201-i data throughout the observation period. The isotopic dataset used in this study was collected during the initial operational period of the G2201-i analyzer, for which we assume that the original factory calibration remained valid throughout the measurement window. This assumption is partly supported by the stability observed in the corresponding CO_2 and CH_4 concentration measurements. Moreover, the focus of this work is on short-term variability, typically over timescales of a few days. Any slow instrumental drift in the isotopic channels would therefore not affect the representativeness of the observed patterns.

2.1.3. Isotopic Analysis

All data analyses were performed using R (version 4.4.2). To investigate temporal variability in isotopic composition, all measurements of CH_4 and CO_2 were aggregated using 10 min averages over the full observation period from 1 March to 10 April 2025. To highlight broader patterns while preserving short-term fluctuations, a moving average with a 36 h window (equivalent to 216 data points) was applied to the time series.

For the Keeling plot analysis, we used ordinary least squares linear regression, with δ values regressed against the inverse of CO_2 concentration. The intercepts of these

https://doi.org/10.3390/environments13030145

regressions were extracted as indicators of the isotopic signature of the dominant source. To quantify the associated uncertainty, we calculated the standard deviation of the intercepts obtained from the individual regression fits.

2.2. Aerosol Analysis

To assess the presence and evolution of Saharan dust intrusions, we integrated aerosol-focused remote sensing observations, trajectory modelling, and in situ measurements. Satellite monitoring relied on the Dust RGB (Red, Green, Blue) product available through EUMETView [49], derived from infrared observations of the Meteosat Second Generation (MSG) [50] SEVIRI instrument (Airbus Defence and Space, Taufkirchen, Germany, IR8.7, IR10.8, and IR12.0). The Dust RGB is a diagnostic false-colour composite in which these three infrared channels are mapped to the red, green, and blue components of the image. This combination enhances the thermal contrast between dust, clouds, and the underlying surface, allowing airborne mineral dust to be visually identified as characteristic pink–magenta areas. This RGB composite is specifically designed to track dust storm dynamics both day and night, providing valuable insight into aerosol transport pathways and atmospheric structure. Vertical information on aerosol layers was obtained from ground-based active remote sensing. Vertical profiles of aerosol optical properties and continuous high-resolution profiles (30 m, 10 s) of aerosol and cloud stratification were analyzed using, respectively, lidar and ceilometer (CL51, Vaisala, Vantaa, Finland) measurements performed within the ACTRIS (Aerosol, Clouds and Trace Gases Research Infrastructure) framework [51] at the CIAO. These observations enabled a detailed characterization of lofted aerosol layers during the dust events. Air mass origin and transport pathways were investigated using NOAA HYSPLIT (Hybrid Single-Particle Lagrangian Integrated Trajectory) backward trajectories [52,53], which supported the identification of source regions and transport patterns associated with each intrusion. Finally, in situ measurements from a nephelometer (Aurora 3000, Ecotech, Knoxfield, VIC, Australia) were used to characterize the scattering properties of mineral dust aerosol, contributing to the assessment of dust event intensity and temporal variability. The nephelometer measured light scattering at three wavelengths (450, 525, and 635 nm), integrating the scattered signal over angular ranges of 9–170° for total scattering and 90–170° for backscattering. Routine calibration was performed using particle-free air for zero checks and CO_2 as the span gas. The dataset was recorded at 1 min resolution, and hourly averages were subsequently calculated. Corrections for truncation and non-Lambertian illumination effects were applied following the methodology described by Muller et al. [54], and all corrections and hourly averaging procedures were implemented using a dedicated Pyton 3.12 script.

3. Results

3.1. The Dust Event

Multiple observational methods confirmed the occurrence of Saharan dust intrusions, revealing three closely spaced events on 10–12 March, 14–16 March, and 22–26 March. The MSG SEVIRI Dust RGB product provided a clear overview of dust outbreaks over the Italian peninsula on 10 March (Figure 1a), 14 March (Figure 1b), and 22 March (Figure 1c).

Complementary cloud remote sensing measurements from the CIAO revealed a persistent aerosol layer on these days, as indicated by the target classification product derived from Cloudnet algorithms (Figure 2a–c). Daily evolution plots are available via the ACTRIS Cloudnet portal [55], offering further insight into the vertical structure and persistence of the aerosol layers.

https://doi.org/10.3390/environments130301

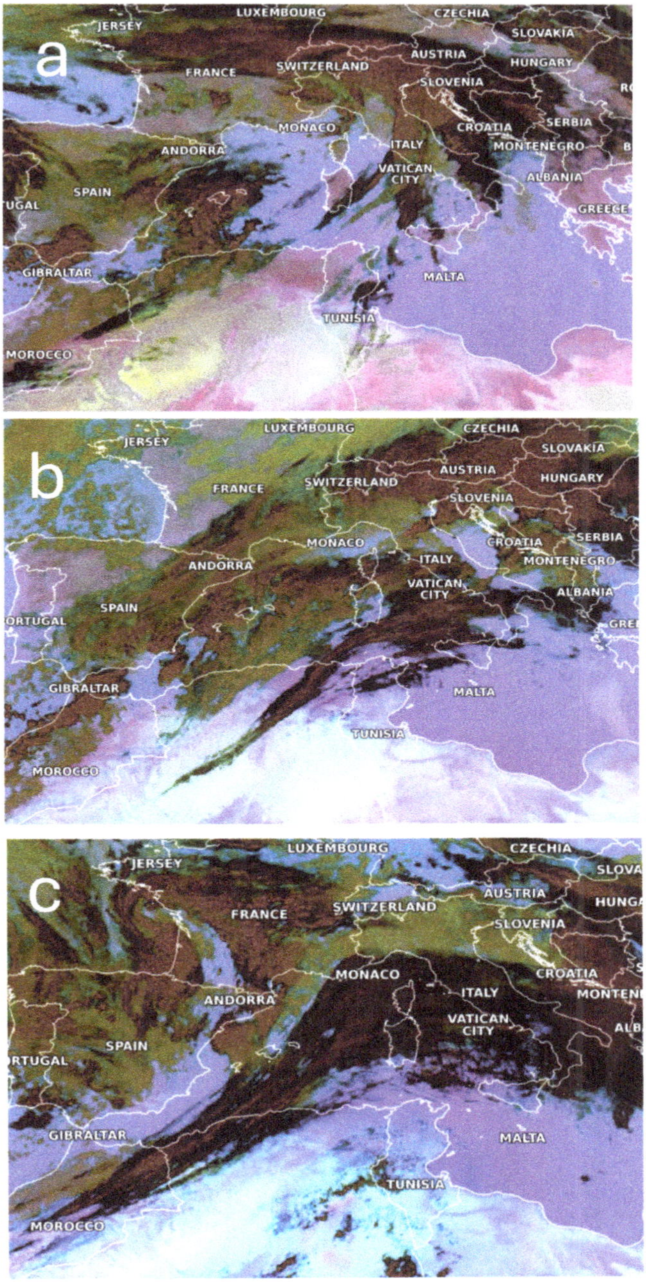

Figure 1. MSG SEVIRI Dust RGB imagery showing the evolution of Saharan dust over the central Mediterranean region. (**a**–**c**) correspond, respectively, to 10 March 2025 at 06:00 UTC, 14 March 2025 at 09:00 UTC, and 22 March 2025 at 12:00 UTC. In the Dust RGB, colours indicate different atmospheric features: magenta—dust or ash clouds; black—cirrus clouds; dark red—thick, high and cold ice clouds; yellow—thick mid-level clouds; dark blue—humid lower-level air; lilac—dry lower-level air. Source: ©EUMETSAT [2025] [49].

 https://doi.org/10.3390/environments13030145

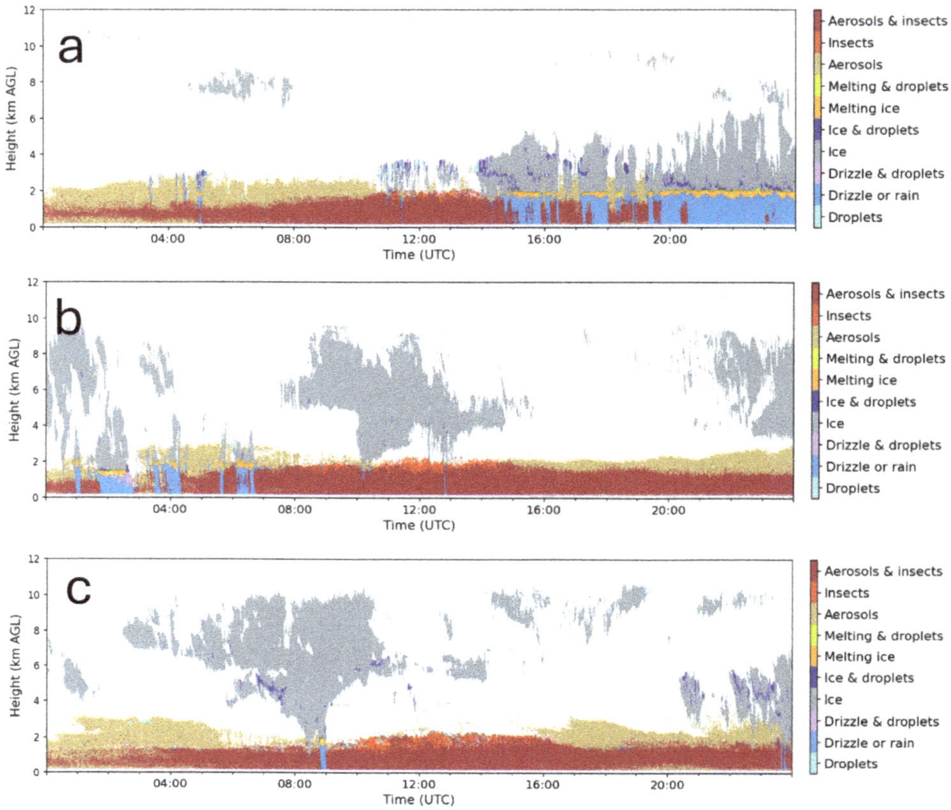

Figure 2. ACTRIS Cloudnet Target classification at Potenza observatory during three Saharan dust episodes. (**a**–**c**) correspond, respectively, to 10 March 2025, 14 March 2025, and 22 March 2025.

These satellite-based and cloud-classification products provide a broad spatial and temporal overview of dust transport across the Mediterranean region. To complement these large-scale observations, ground-based lidar measurements from the ACTRIS/EARLINET network [56] were examined. These measurements were acquired using the Potenza Lidar for Particle Observation (POLPO) which operates at 355 nm, 532 nm and 1064 nm, at CIAO, and is designed for continuous monitoring of aerosol layers. The vertical profiles of aerosol optical properties measured on 15 March 2025 between 20:36 and 21:36 UTC (Figure 3a) indicate the presence of dust-contaminated air masses within the planetary boundary layer, as evidenced by depolarization ratios at 532 nm ranging between 0.1 and 0.2. The lidar profiles acquired on 21 March 2025 between 20:19 and 21:14 UTC (Figure 3b) reveal similar optical characteristics, confirming the presence of Saharan dust within the lower troposphere. In this case, the dust layer is more vertically extended, with a pronounced signal between an altitude of 2 and 3 km. The depolarization ratio at 532 nm exceeds 0.3 throughout this layer, indicating the dominance of pure mineral dust particles. Due to the persistent rainfall during the analyzed period, most lidar observations were not available. For this reason, all other days were excluded from analysis.

https://doi.org/10.3390/environments13030

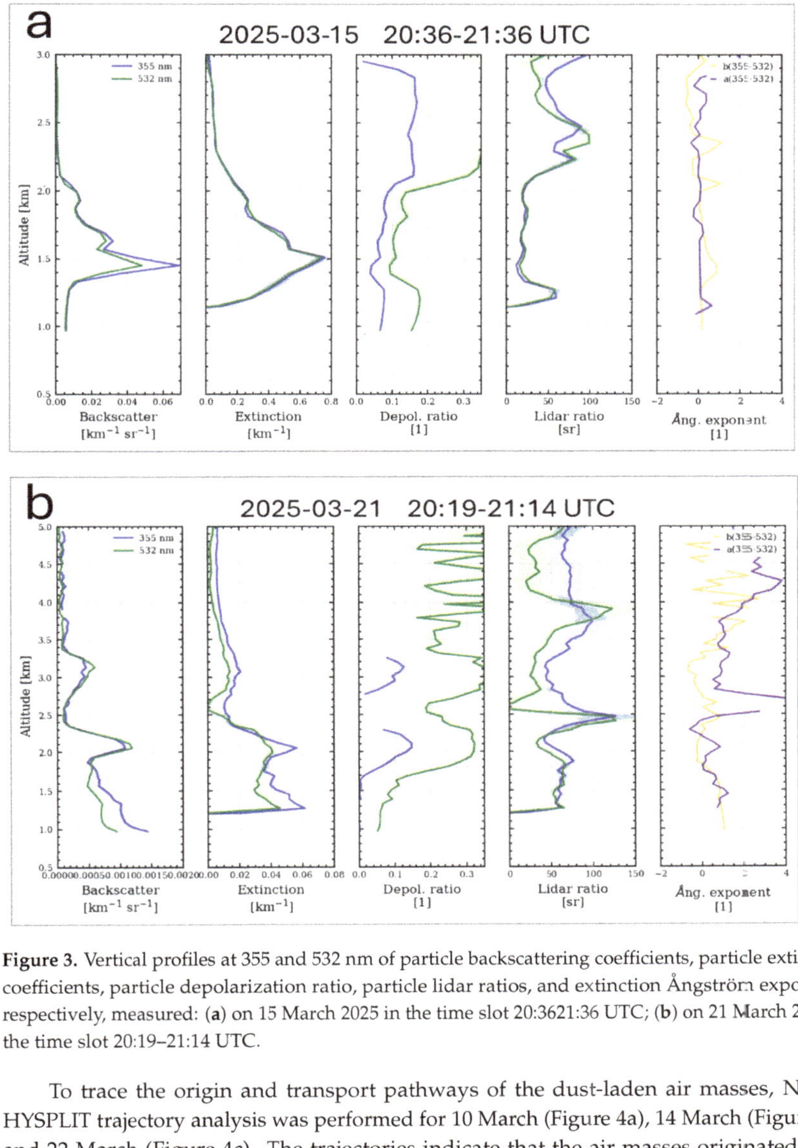

Figure 3. Vertical profiles at 355 and 532 nm of particle backscattering coefficients, particle extinction coefficients, particle depolarization ratio, particle lidar ratios, and extinction Ångström exponents, respectively, measured: (**a**) on 15 March 2025 in the time slot 20:3621:36 UTC; (**b**) on 21 March 2025 in the time slot 20:19–21:14 UTC.

To trace the origin and transport pathways of the dust-laden air masses, NOAA HYSPLIT trajectory analysis was performed for 10 March (Figure 4a), 14 March (Figure 4b), and 22 March (Figure 4c). The trajectories indicate that the air masses originated from North Africa, following distinct paths. During the first intrusion (10 March), air parcels originated from central North Africa and traveled over Libya, Algeria, and Tunisia before reaching southern Italy. The second event (14 March) was characterized by a more westerly trajectory, with air masses primarily originating from Morocco and crossing the western Mediterranean Basin. In the third event (22 March), the air parcels again passed through Libya and Tunisia, with additional contributions from northern Europe, suggesting a more complex and layered transport pattern.

https://doi.org/10.3390/environments13030145

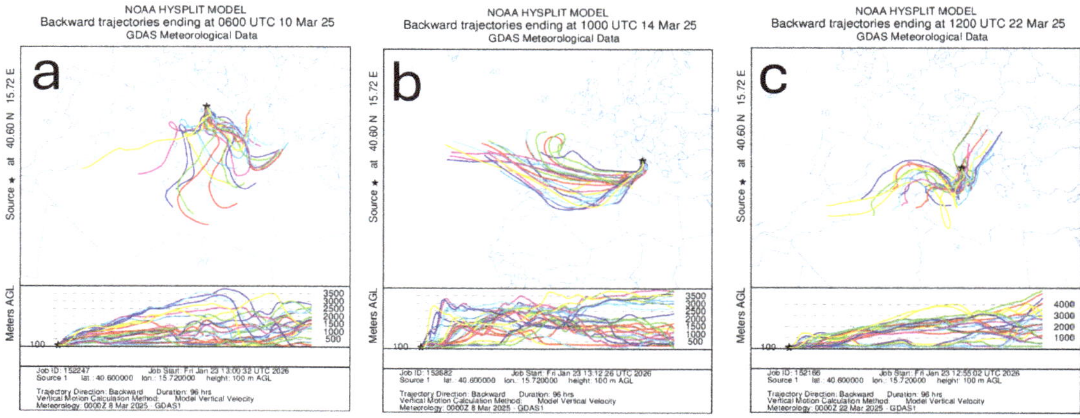

Figure 4. HYSPLIT 96 h back trajectories for POT station. (**a**) Backward trajectories ending at 06:00 UTC 10 March 2025, (**b**) Backward trajectories ending at 10:00 UTC 14 March 2025, and (**c**) Backward trajectories ending at 12:00 UTC 22 March 2025.

The nephelometer observations (Figure 5a) show a pronounced shift in the spectral behavior of light scattering. Before the dust intrusion, the scattering coefficients exhibited the characteristic pattern of fine aerosol, with σsp (635 nm) < σsp (525 nm) < σsp (450 nm), resulting in positive scattering Ångström exponents (SAE) values around 1. During the dust events, this ordering reversed: σsp (635 nm) became the dominant wavelength, driving the SAE to values well below 1 and reaching −0.5 (Figure 5b). Such an inversion is characteristic of coarse mineral particles whose diameters are comparable to or larger than the measurements of wavelengths, thereby enhancing scattering at longer wavelengths. During the first event, a modest increase in σsp at 635 nm between 10 and 12 March, relative to pre-dust conditions, already points to the early arrival of coarse-mode particles. The second event displays a sharper and more intense peak at longer wavelength between 14 and 16 March, pointing to a more concentrated intrusion. This episode is marked by a sudden and pronounced increase in particulate concentration, producing a well-defined scattering signal. In contrast, the third event shows a broader and more prolonged enhancement in long-wavelength scattering, consistent with a prolonged dust presence over the region from 22 to 26 March. The persistence of elevated scattering values indicates a stable dust-laden air mass influencing the measurements over several days.

3.2. *Temporal Variability of Isotopic Composition*

Figure 6a,b present the evolution of $\delta^{13}C–CH_4$ and CH_4 mole fractions, respectively. The time series reveals several distinct phases, including periods of relative stability and intervals marked by abrupt changes.

Figure 6a shows alternating enrichment and depletion in $\delta^{13}C–CH_4$ values, with transitions almost aligned with the timing of Saharan dust intrusions (10–26 March). Figure 6b shows the corresponding CH_4 mole fractions, characterized by an overall trend with localized fluctuations. A clear drop occurs during the first two dust intrusion events, which exhibit very similar pattern.

https://doi.org/10.3390/environments13030

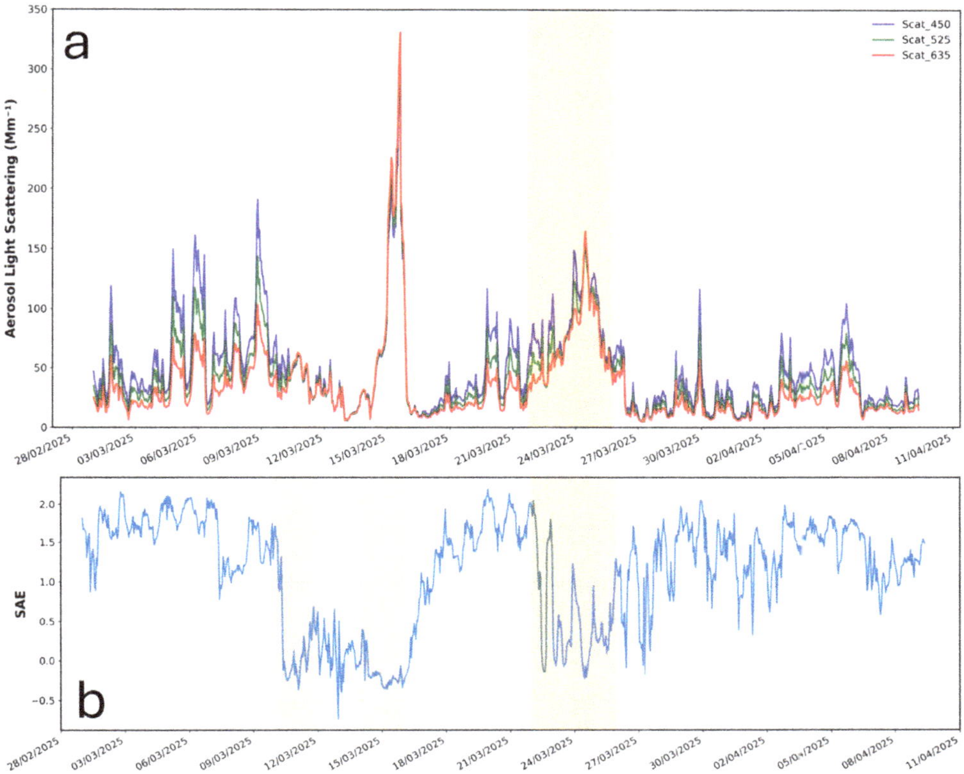

Figure 5. Time series of (**a**) aerosol light scattering (Mm^{-1}) and (**b**) the scattering Ångström exponent ($SAE_{450–635}$) measured by the Aurora 3000-Ecotech nephelometer during the study period. Shaded intervals highlight the temporal windows dominated by coarse aerosol presence: 10–12 March, 14–16 March, and 22–26 March.

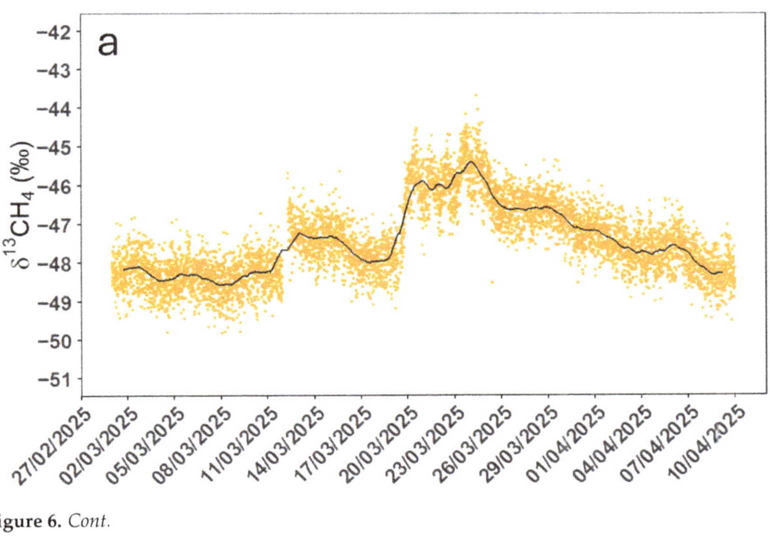

Figure 6. *Cont.*

https://doi.org/10.3390/environments13030145

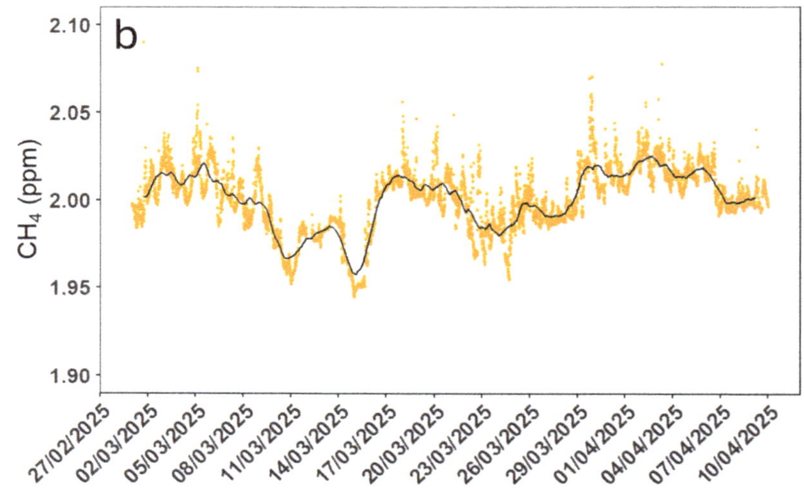

Figure 6. 10 min blocks covering the period from 1 March to 10 April. The black lines represent the 36 h moving average. Shaded intervals indicate three Saharan dust intrusion events: 10–12 March, 14–16 March, and 22–26 March. Figure (**a**) reports the High-Precision (HP) $\delta^{13}C\text{-}CH_4$ values (‰ VPDB), and figure (**b**) shows the corresponding CH_4 mole fractions (ppm).

Figure 7a,b illustrate the behavior of $\delta^{13}C\text{–}CO_2$ and CO_2 mole fractions, respectively. The $\delta^{13}C\text{–}CO_2$ time series shows a gradual increase over the observed period, as expected during the early growing season, with a clear inflection occurring between 11 and 23 March. CO_2 mole fractions remain relatively stable overall, exhibiting only slight increases beginning on 15 and again on 21 March.

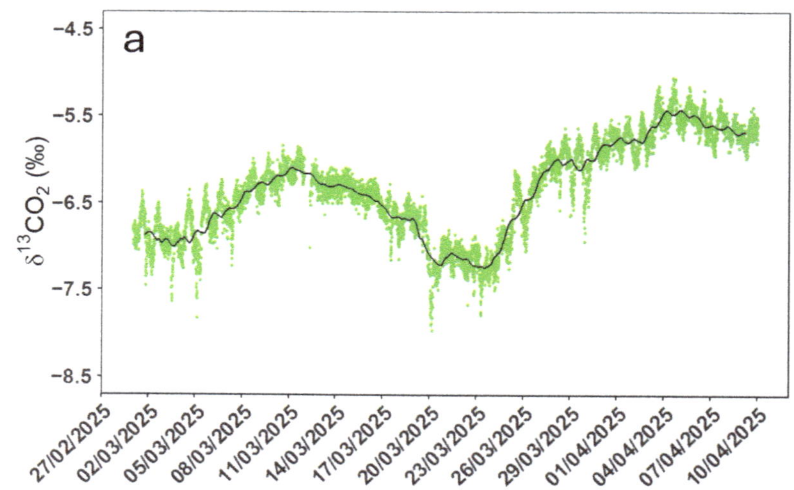

Figure 7. *Cont.*

https://doi.org/10.3390/environments13030

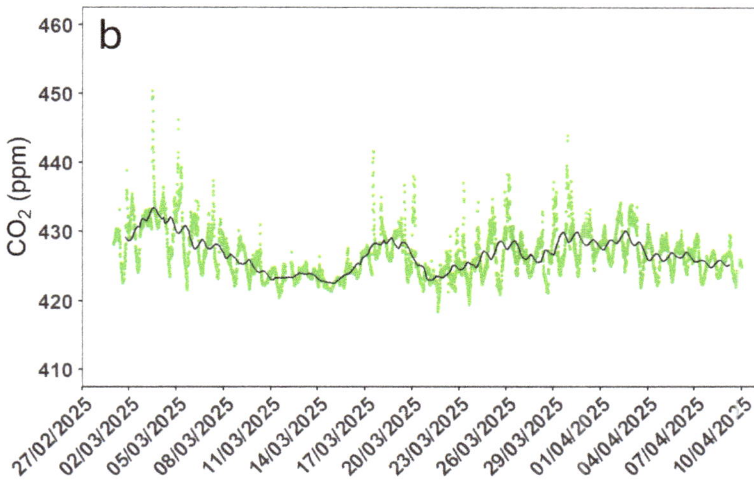

Figure 7. 10 min blocks covering the period from 1 March to 10 April. The black lines represent the 36 h moving average. Shaded intervals indicate three Saharan dust intrusion events: 10–12 March, 14–16 March, and 22–26 March. Figure (**a**) reports the High Precision (HP) $\delta^{13}C$- CO_2 values (‰ VPDB), and figure (**b**) shows the corresponding CO_2 mole fractions (ppm).

Together, these four figures provide a comprehensive overview of the dynamics of stable carbon isotopes and GHG concentrations before, during, and after the Saharan dust events. The patterns observed will be further examined in the discussion section to explore potential mechanisms and implications.

3.2.1. Keeling Plots of CO_2

To investigate the isotopic behavior of atmospheric CO_2 during Saharan dust intrusions, Keeling plots [35] (Figure 8a–c) were constructed for three distinct time windows corresponding to the periods with the highest R^2 values, in order to maximize the reliability of the inferred source signature. Each plot represents a linear regression between $\delta^{13}C$–CO_2 and the inverse of CO_2 concentration. The regression slopes and intercepts vary across the three intervals, reflecting differences in the isotopic composition of CO_2 and its sources.

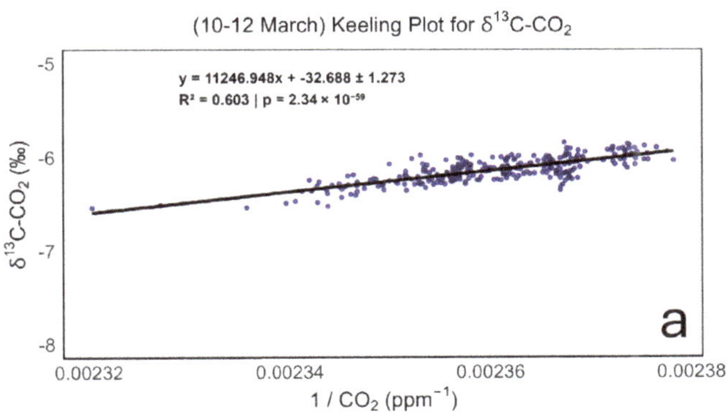

Figure 8. *Cont.*

https://doi.org/10.3390/environments13030145

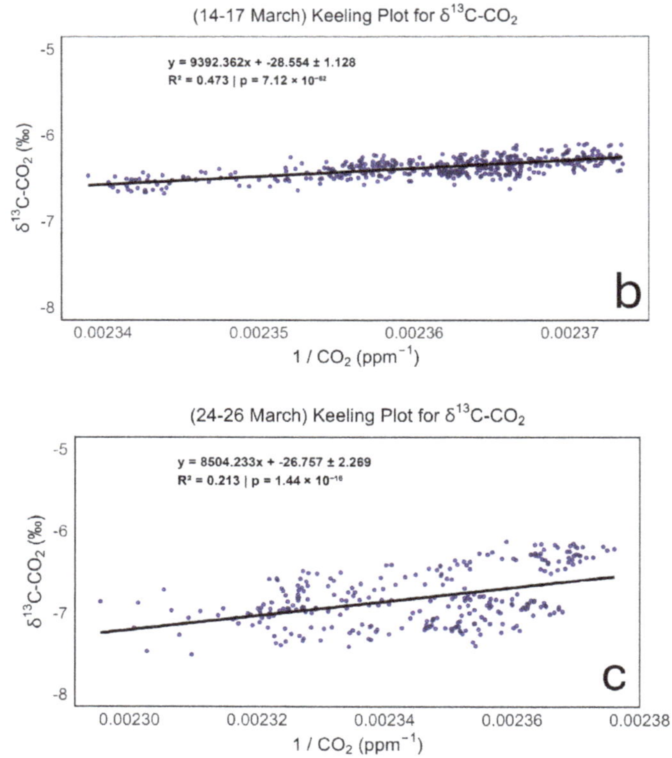

Figure 8. Keeling plots of $\delta^{13}C–CO_2$ versus the inverse of CO_2 concentration $1/CO_2$ for three distinct Saharan dust intrusion periods, based on 10 min blocks: (**a**) 10–12 March, (**b**) 14–17 March, and (**c**) 24–26 March 2025. The intercepts, shown with their standard deviations, indicate the isotopic fingerprint of the emission source. The p-value associated with the regression is reported as **p**.

The first event (10–12 March) shows a relatively strong correlation ($R^2 = 0.603$), suggesting a coherent isotopic signal. The second event (14–17 March) presents a slightly lower R^2 (0.473), yet still indicates a consistent isotopic trend, possibly reflecting enhanced heterogeneity in air mass composition. The third event (24–26 March) exhibits the weakest correlation ($R^2 = 0.213$), which may be attributed to more stratified transport regimes and mixed air masses, as suggested by trajectory analysis.

3.2.2. Keeling Plots of CH_4

To explore the isotopic behavior of atmospheric CH_4 during Saharan dust intrusions, Keeling plots [36] (Figure 9a–c) were constructed for the same three time windows used in the CO_2 analysis: 10–12 March, 14–17 March, and 24–26 March 2025. Each plot represents a linear regression between $\delta^{13}C–CH_4$ and the inverse of CH_4 concentration. The results show a very weak correlation, with low R^2 values suggesting that CH_4 variability during these events is influenced by multiple overlapping sources and does not conform to the assumptions of a single-source mixing model. Nevertheless, the consistently very negative intercepts fall within the characteristic range of biogenic CH_4, suggesting that background CH_4 during the dust intrusions was dominated by biogenic emissions.

https://doi.org/10.3390/environments13030

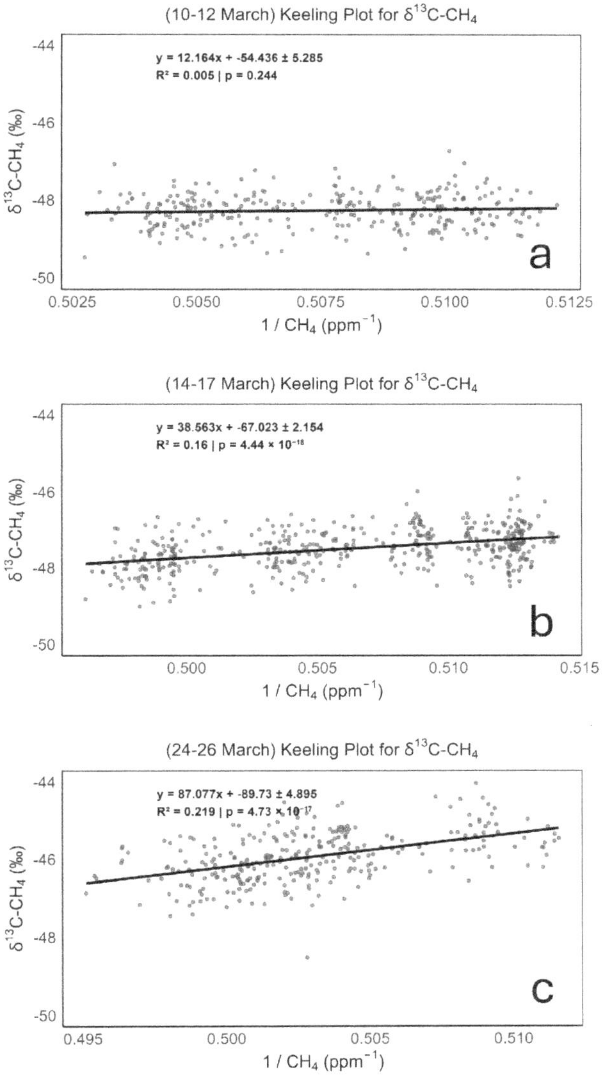

Figure 9. Keeling plots of $\delta^{13}C$–CH_4 versus the inverse of CH_4 concentration $1/CH_4$ for three distinct Saharan dust intrusion periods, based on 10 min blocks: (**a**) 10–12 March, (**b**) 14–17 March, and (c) 24–26 March 2025. The *p*-value associated with the regression is reported as **p**.

4. Discussion

The integrated analysis, combining satellite and ground-based remote sensing observations, in situ aerosol optical measurements, air mass trajectory modeling, and continuous monitoring of GHG stable isotopes, reveals a coherent link between Saharan dust intrusions and perturbations in regional carbon dynamics. Although the temporal correspondence between the dust intrusions and the observed isotopic shifts is evident, the alignment is not perfectly synchronous. The physical presence of mineral aerosol and the chemical processes influencing $\delta^{13}C$-CH_4 and $\delta^{13}C$-CO_2 can evolve on different timescales. Gas-phase

 https://doi.org/10.3390/environments13030145

oxidation may exhibit a time lag relative to the arrival of dust-laden air masses, and such reactions can continue or intensify even after the peak aerosol load has passed, leading to delayed isotopic responses. Additionally, part of this mismatch may arise from the photochemical nature of these processes, which operate under sufficient solar radiation. Consequently, the timing of isotopic changes along the air mass trajectory may carry additional uncertainty, reflecting the variable onset of photochemically driven reactions. In this context, the shift toward less negative $\delta^{13}C\text{-}CH_4$ values around 20 March likely reflects the delayed and combined influence of two closely spaced Saharan dust intrusions. Although the aerosol peaks occurred earlier, the associated changes in oxidative capacity can persist and partially overlap, producing an integrated isotopic signal that does not coincide exactly with the timing of maximum dust load.

In situ measurements of aerosol optical properties revealed a marked contribution from coarse mineral particles, a defining feature of Saharan dust intrusions, indicating that the sampled air masses were influenced by dust-laden conditions. The first and third episodes were characterized by slightly enhanced scattering coefficients at longer wavelengths, displaying less sharply defined peaks that suggest less intense dust plumes. In contrast, the second intrusion exhibited a markedly higher scattering signal at longer wavelengths, with a narrow and well-defined peak of short duration, indicative of a more concentrated and intense dust transport episode. Enhancements in aerosol scattering were accompanied by notable changes in particle size distribution, as reflected by the scattering Ångström exponent (SAE). During the first two events, SAE values approached zero, consistent with the dominance of coarse mineral particles. In contrast, the third event showed a less pronounced and more variable decrease in SAE, suggesting a more heterogeneous aerosol composition. This interpretation is consistent with the back trajectory analysis: while the first two intrusions were clearly linked to air masses originating from North African source regions, the third event displayed a more complex transport pattern. In this case, the air parcels travelled over Libya and Tunisia but also included contributions from northern Europe, indicating the arrival of mixed and layered air masses during the 22 March intrusion. Such variability in aerosol origin may also help explain the different isotopic response observed in $\delta^{13}C\text{–}CH_4$ during this period, as air parcels with distinct source regions and transport histories can carry chemically diverse mixtures of dust and reactive gases. Notably, the intense scattering peak observed on 14–16 March coincides with the most pronounced subsequent enrichment in $\delta^{13}C\text{-}CH_4$, which becomes evident after 17 March. This temporal sequence suggests that the conditions associated with the second intrusion were particularly conducive to promoting CH_4 isotopic fractionation.

Specifically, the observed decrease in CH_4 mole fractions accompanied by an enrichment in $\delta^{13}C\text{–}CH_4$ points to isotopic fractionation likely driven by enhanced oxidative reactions. In addition to this, diurnal variability provides further insight into the short-term dynamics of the system. Previous work by Buono et al. [31] has characterized the expected seasonal behaviour of CO_2 and $\delta^{13}C\text{-}CO_2$ at the site, and over the period considered here (1 March–20 April) both species exhibit patterns consistent with those established seasonal cycles. For methane, the March diurnal cycle shows a more pronounced daytime decrease in CH_4 mole fractions. In contrast, $\delta^{13}C\text{-}CH_4$ in March remains systematically less negative than in April, with this behaviour persisting throughout most of the month. These diurnal features, although not the primary focus of the present study, are consistent with the interpretation that oxidation processes may modulate the observed isotopic signatures.

One plausible mechanism involves the catalytic role of Fe-rich mineral dust in generating reactive Cl radicals (Cl•), which increases the oxidative capacity of the atmosphere and preferentially removes lighter CH_4 isotopes. Simultaneously, a light increase in atmo-

 https://doi.org/10.3390/environments1303C

spheric CO_2 concentrations was observed during the dust events. This rise is plausibly associated with a temporary reduction in photosynthetic activity caused by diminished incoming solar radiation on the surface, itself a consequence of the elevated aerosol loads documented from vertical profiles of aerosol optical properties over the observation period. Correspondingly, $\delta^{13}C$–CO_2 values show a noticeable decline, consistent with reduced uptake of $^{12}CO_2$ by plants. Since photosynthesis preferentially fixes the lighter isotope (^{12}C), diminished photosynthetic activity leads to less ^{12}C being removed from the atmosphere. As a result, the $\delta^{13}C$–CO_2 values temporarily decline, reflecting the transient decrease in photosynthetic carbon uptake. Outside of dust intrusion periods (10–26 March 2025), the $\delta^{13}C$–CO_2 trend displays a progressive increase, as illustrated in the corresponding figures, which is consistent with the expected seasonal enhancement of photosynthetic activity during this time of year [57].

To further investigate the isotopic source signatures associated with the dust-intrusion periods, Keeling plots were generated for CO_2 and CH_4 across the three restricted time windows. These analyses offer additional insight into mixing processes and potential source contributions, influencing the observed isotopic variability. While the CO_2 Keeling plots display clearer linear relationships, the CH_4 plots consistently show very weak correlations, with low coefficients of determination. This contrast reflects the more complex and regionally integrated atmospheric behavior of CH_4, which rarely conforms to a simple two-end-member mixing framework on which the Keeling plot approach relies [36]. Moreover, the markedly different atmospheric abundances of the two gases further contribute to this divergence: CO_2 is over two orders of magnitude more abundant than CH_4, meaning that small measurement uncertainties propagate very differently. As a result, the CH_4 Keeling plots are inherently more sensitive to noise and variability, reinforcing the weaker and less interpretable linear relationships observed during the dust-intrusion periods. Although the weak correlations observed in the CH_4 Keeling plots limit the robustness of source apportionment, the consistently very negative intercepts remain compatible with a predominantly biogenic CH_4 signature. This interpretation is also consistent with the well-documented [58,59] association between Saharan dust and long-range transport of microbial and organic material, which may contribute to the complex mixture of bio-mineral particles influencing the isotopic composition of the sampled air masses. The Keeling intercepts obtained for CO_2 (−32.7‰, −28.5‰, and −26.8‰) fall within the typical range of C_3 ecosystem respiration [60] and indicate that, during the Saharan dust intrusions, the isotopic signature of CO_2 was dominated by local biogenic sources. The dust intrusions do not introduce a characteristic CO_2 isotopic signature of their own but instead modify atmospheric mixing and surface processes, thereby influencing the apparent intercept.

Marteen et al. [19] proposed, also, that the MDSA mechanism plays a pivotal role in regional-scale atmospheric chemistry, potentially driving ozone (O_3) suppression, VOC oxidation, secondary organic aerosol formation, and HCl production through heterogeneous reactions involving Cl. Furthermore, Bonasoni et al. [61] investigated the impact of Saharan dust intrusions on tropospheric O_3 levels at Mt. Cimone station in Northern Italy. During 12 documented dust transport events between June and December 2000, they observed a pronounced anticorrelation between coarse particle concentrations and surface O_3 mixing ratio. Specifically, southerly flows carrying mineral dust from North Africa were associated with significant O_3 reductions compared to background free-tropospheric conditions. These findings suggest that elevated concentrations of mineral aerosol can facilitate heterogeneous O_3 destruction on particle surfaces, even at considerable distances from the dust source regions.

https://doi.org/10.3390/environments13030145

Taken together, the process-based hypothesis of Marteen et al. [18] and the observational evidence from Bonasoni et al. [61] provide a coherent framework for interpreting the O_3 depletion events observed at Mt. Cimone. In our study, the isotopic shifts in CH_4 and CO_2 observed during Saharan dust intrusion further support the occurrence of dust-induced chemical processing, likely involving Cl-mediated heterogeneous reactions consistent with the MDSA mechanism. This convergence of chemical theory, observational data, and isotopic evidence reinforces the role of mineral dust as a key modulator of tropospheric oxidizing capacity in the Mediterranean region.

5. Conclusions

This study provides preliminary evidence that Saharan dust intrusions can alter the isotopic composition of atmospheric methane (CH_4) and carbon dioxide (CO_2), although through distinct and partly independent mechanisms. The combined analysis of aerosol optical properties, air mass trajectories, and continuous isotopic measurements shows that dust-laden air masses can affect regional carbon dynamics by modifying both the physical environment and the chemical reactivity of the atmosphere. For CH_4, the observed decreases in CH_4 mole fractions accompanied by enrichments in $\delta^{13}C–CH_4$ during dust events suggest enhanced isotopic fractionation. Although CH_4 Keeling plots did not yield strong linear relationships, reflecting the complex, multi-source nature of atmospheric CH_4, the isotopic time series nonetheless indicates that dust intrusions can modulate CH_4 isotopic composition. For CO_2, the temporary increases in CO_2 concentrations and concurrent declines in $\delta^{13}C–CO_2$ are consistent with a short-term reduction in photosynthetic uptake driven by dust-induced radiative suppression. In contrast to CH_4, CO_2 Keeling plots displayed clearer mixing relationships, highlighting the stronger influence of local processes on short-term CO_2 variability during dust events. Taken together, these findings demonstrate that Saharan dust intrusions exert a measurable influence on the isotopic composition of both CH_4 and CO_2, acting through a combination of radiative, chemical, and transport-related processes. Furthermore, the alignment with O_3 depletion patterns documented in previous studies reinforces the broader role of mineral dust in modulating tropospheric oxidative capacity across the Mediterranean basin. However, given the complexity of the processes involved and the limited temporal resolution of the events examined, these interpretations should be considered indicative rather than conclusive. Future research should include extended observational campaigns and targeted in situ chemical analyses to fully characterize the mechanisms driving aerosol–gas coupling and their broader climatic implications.

Author Contributions: Conceptualization, I.Z.; methodology, I.Z., A.B. and F.D.; software, I.Z., E.L. and C.C.; validation, I.Z., A.B., T.L., B.D.R. and F.D.; formal analysis, I.Z., A.B., B.D.R., T.L. and F.D.; investigation, I.Z., A.B., F.D., E.L., F.C., T.L., M.R., B.D.R. and D.S.; resources, D.A., E.R. and A.G.; data curation, I.Z., A.B., F.D., T.L. and B.D.R., writing—original draft preparation, I.Z. and A.B.; writing—review and editing, I.Z., A.B., F.D., E.L., F.C., T.L., D.A., C.C., G.D.F., A.G., G.M., G.G., A.G.d.S., C.R.C., S.T., B.D.R., M.R., D.S. and L.M.; visualization, I.Z., A.B., B.D.R., T.L., F.D., E.L. and F.C.; supervision, A.G.d.S., C.R.C., S.T., G.M. and L.M.; funding acquisition, G.D.F., S.T. and L.M. All authors have read and agreed to the published version of the manuscript.

Funding: This research was funded by MUR (Italian Ministry of University and Research) through PON "Ricerca e Innovazione 2014–2020" under the following projects: PRO ICOS MED "Potenziamento della Rete di Osservazione ICOS-Italia nel Mediterraneo", contract PIR01_00019, CUP B85D18000340001 (7 June 2019–18 February 2022), PRO ICOS MED (Potenziamento della Rete di Osservazione ICOS-Italia nel Mediterraneo—Rafforzamento del capitale umano)—Avviso MUR D.D. n. 2595 del 24 December 2019 Piano Stralcio "Ricerca e Innovazione 2015–2017", CIR01_00019,

https://doi.org/10.3390/environments13030

CUP B58I20000210001, PER-ACTRIS-IT (Potenziamento della componente italiana della Infrastruttura di Ricerca Aerosol, Clouds, and Trace Gases Research Infrastructure)–Avviso MUR D.D. n. 2595 del 24.12.2019 Piano Stralcio "Ricerca e Innovazione 2015–2017", PIR01_C0015, CUP B17E19000000007, ITINERIS, Italian Integrated Environmental Research Infrastructure System (IR0000032, D.D. n.130/2022—CUPB53C22002150006) Funded by EU—Next Generation EU PNRR—Mission 4—Component 2—Investment 3.1 and Ordinary Fund for Research Institutions and Organizations (FOE), funded by the Italian Ministry of University and Research (D.M. n. 744, 8 October 2020).

Institutional Review Board Statement: Not applicable.

Informed Consent Statement: Not applicable.

Data Availability Statement: The dataset featuring 10 min aggregated blocks of stable carbon isotope data ($\delta^{13}C$–CO_2, $\delta^{13}C$–CH4) and the respective mole fractions of CO_2 and CH_4 is available via the ITINERIS HUB [40].

Acknowledgments: We are grateful to ACTRIS and the Finnish Meteorological Institute for providing the ceilometer dataset and the ACTRIS-EARLINET network for providing aerosol optical parameters essential for reconstructing and analyzing the vertical profiles of atmospheric aerosols. We also acknowledge the Joint Research Unit "ICOS Italia", funded by Ministry of University and Research. We acknowledge the support of EUMETSAT for providing access to the EUMETView platform. Our appreciation also goes to the NOAA Air Resources Laboratory (ARL) for the provision of the HYSPLIT transport and dispersion model and the READY website which have been fundamental in tracking atmospheric transport processes.

Conflicts of Interest: The authors declare no conflicts of interest.

ferences

Prather, M.J.; Holmes, C.D.; Hsu, J. Reactive greenhouse gas scenarios: Systematic exploration of uncertainties and the role of atmospheric chemistry. *Geophys. Res. Lett.* **2012**, *39*, L09803. [CrossRef]

IPCC: 2019 Refinement to the 2006 IPCC Guidelines for National Greenhouse Gas Inventories—IPCC. Available online: https://www.ipcc.ch/report/2019-refinement-to-the-2006-ipcc-guidelines-for-national-greenhouse-gas-inventories/ (accessed on 20 December 2025).

Schaefer, H. On the Causes and Consequences of Recent Trends in Atmospheric Methane. *Curr. Clim. Change Rep.* **2019**, *5*, 259–274. [CrossRef]

Nisbet, E.G.; Manning, M.R. What is causing the methane surge? Atmospheric methane's recent growth has been driven by biogenic sources and COVID-19 impacts. *Science* **2026**, *391*, 556–557. [CrossRef] [PubMed]

Drinkwater, A.; Palmer, P.I.; Feng, L.; Arnold, T.; Lan, X.; Michel, S.E.; Parker, R.; Boesch, H. Atmospheric data support a multi-decadal shift in the global methane budget towards natural tropical emissions. *Atmos. Chem. Phys.* **2023**, *23*, 8429–8452. [CrossRef]

Kirschke, S.; Bousquet, P.; Ciais, P.; Saunois, M.; Canadel, J.G.; Dlugokencky, E.J.; Bergamaschi, P.; Bergmann, D.; Blake, D.R.; Bruhwiler, L.; et al. Three decades of global methane sources and sinks. *Nat. Geosci.* **2013**, *6*, 813–823. [CrossRef]

Nisbet, E.G.; Manning, M.R.; Dlugokencky, E.J.; Fisher, R.E.; Lowry, D.; Michel, S.E.; Myhre, C.L.; Platt, S.M.; Allen, G.; Bousquet, P.; et al. Very Strong Atmospheric Methane Growth in the 4 Years 2014–2017: Implications for the Paris Agreement. *Glob. Biogeochem. Cycles* **2019**, *33*, 318–342. [CrossRef]

Lawler, M.J.; Sander, R.E.; Carpenter, L.; Lee, J.D.; von Glasow, R.; Sommariva, R.; Saltzman, E.S. HOCl and Cl_2 observations in marine air. *Atmos. Chem. Phys.* **2011**, *11*, 7617–7628. [CrossRef]

Hossaini, R.; Chipperfield, M.P.; Saiz-Lopez, A.; Fernandez, R.; Monks, S.; Feng, W.P.; Brauer, P.; von Glasow, R. A global model of tropospheric chlorine chemistry: Organic versus inorganic sources and impact on methane oxidation. *J. Geophys. Res. Atmos.* **2016**, *121*, 14271–14297. [CrossRef]

Gromov, S.; Brenninkmeijer, C.A.M.; Jöckel, P.A. Very limited role of tropospheric chlorine as a sink of the greenhouse gas methane. *Atmos. Chem. Phys.* **2018**, *18*, 9831–9843. [CrossRef]

Hanson, R.S.; Hanson, T.E. Methanotrophic bacteria. *Microbiol. Rev.* **1996**, *60*, 439–471. [CrossRef] [PubMed]

Saueressig, G.; Bergamaschi, P.; Crowley, J.N.; Fischer, H.; Harris, G.W. Carbon kinetic isotope effect in the reaction of CH_4 with Cl atoms. *J. Geophys. Res. Atmos.* **1995**, *22*, 1225–1228. [CrossRef]

https://doi.org/10.3390/environments13030145

13. Lan, X.; Nisbet, E.G.; Dlugokencky, E.J.; Michel, S.E. Improved Constraints on Global Methane Emissions and Sinks Us $\delta^{13}C$-CH_4. *J. Geophys. Res. Atmos.* **2021**, *35*, e2021GB007000. [CrossRef] [PubMed]
14. Quay, P.D.; King, S.L.; Stusman, J.; Wilbur, D.O.; Steele, L.P.; Fung, I.; Gammon, R.H.; Brown, T.A.; Farwell, G.W.; Grootes, P. et al. Carbon isotopic composition of atmospheric CH_4: Fossil and biomass burning source strengths. *Glob. Biogeochem. Cy* **1991**, *5*, 25–47. [CrossRef]
15. Saunois, M.; Stavert, A.R.; Poulter, B.; Bousquet, P.; Canadell, J.G.; Jackson, R.B.; Raymond, P.A.; Dlugokencky, E.J.; Houweling Patra, P.K.; et al. The Global Methane Budget 2000–2017. *Earth Syst. Sci. Data* **2020**, *12*, 1561–1623. [CrossRef]
16. Basu, S.; Lan, X.; Dlugokencky, E.; Michel, S.; Schwietzke, S.; Miller, J.B.; Bruhwiler, L.; Oh, Y.; Tans, P.P.; Apadula, F.; et Estimating emissions of methane consistent with atmospheric measurements of methane and $\delta^{13}C$ of methane. *Atmos. Ch Phys.* **2022**, *22*, 15351–15377. [CrossRef]
17. Cheewaphongphan, P.; Chatani, S.; Saigusa, N. Exploring Gaps between Bottom-Up and Top-Down Emission Estimates Based Uncertainties in Multiple Emission Inventories: A Case Study on CH_4 Emissions in China. *Sustainability* **2019**, *11*, 2054. [CrossR
18. Röckmann, T.; van Herpen, M.; Brashear, C.; van der Veen, C.; Gromov, S.; Li, Q.; Saiz-Lopez, A.; Meidan, D.; Barreto, A.; Pr N.; et al. The use of $\delta^{13}C$ in CO to determine removal of CH_4 by Cl radicals in the atmosphere. *Environ. Res. Lett.* **2024**, *19*, 064 [CrossRef]
19. Maarten, M.J.; van Herpena, W.; Qinyi, L.; Saiz-Lopezb, A.; Liisbergc, J.B.; Röckmann, T.; Cuevas, C.A.; Fernandez, R.P.; Mak, J Mahowald, N.M.; et al. Photocatalytic chlorine atom production on mineral dust–sea spray aerosols over the North Atlantic. P *Natl. Acad. Sci. USA* **2023**, *120*, e2303974120. [CrossRef]
20. Mona, L.; Amiridis, V.; Cuevas, E.; Gkikas, A.; Trippetta, S.; Vandenbussche, S.; Benedetti, A.; Dagsson-Waldhauserova, Formenti, P.; Haefele, A.; et al. Observing Mineral Dust in Northern Africa, the Middle East, and Europe: Current Capabilit and Challenges ahead for the Development of Dust Services. *Bull. Am. Meteorol. Soc.* **2023**, *104*, E2223–E2264. [CrossRef]
21. Brass, M.; Röckmann, T. Continuous-flow isotope ratio mass spectrometry method for carbon and hydrogen isotope measureme on atmospheric methane. *Atmos. Meas. Tech.* **2010**, *3*, 1707–1721. [CrossRef]
22. Zhang, Y.; Sunwoo, Y.; Kotamarthi, V.; Carmichael, G.R. Photochemical oxidant processes in the presence of dust: An evalua of the impact of dust on particulate nitrate and processes ozone formation. *J. Appl. Met.* **1994**, *33*, 813–824. [CrossRef]
23. Allan, W.; Struthers, H.; Lowe, D.C. Methane carbon isotope effects caused by atomic chlorine in the marine boundary la Global model results compared with Southern Hemisphere measurements. *J. Geophys. Res. Atmos.* **2007**, *112*, D04306. [CrossR
24. Qinyi, L.; Fernandez, R.P.; Hossaini, R.; Iglesias-Suarez, F.; Cuevas, C.A.; Apel, E.C.; Kinnison, D.E.; Lamarque, J.-F.; Saiz-Lop A. Reactive halogens increase the global methane lifetime and radiative forcing in the 21st century. *Nat. Commun.* **2022**, *13*, 2 [CrossRef]
25. Arimoto, R. Eolian dust and climate: Relationships to sources, tropospheric chemistry, transport and deposition. *Earth Sci.* **2001**, *54*, 29–42. [CrossRef]
26. Cantrell, C.A.; Shetter, R.E.; McDaniel, A.H.; Calvert, J.G.; Davidson, J.A.; Lowe, D.C. Carbon kinetic isotope effect in the oxidat of methane by the hydroxyl radical. *J. Geophys. Res. Atmos.* **1990**, *95*, 22455–22462. [CrossRef]
27. Bergamaschi, P.; Bruhl, C.; Brenninkmeijer, C.A.M.; Saueressig, G.; Crowley, J.N.; Groos, J.U.; Fischer, H.; Crutzen, P.J. Implicati of the large carbon kinetic isotope effect in the reaction CH_4^+ Cl for the $^{13}C/^{12}C$ ratio of stratospheric CH_4. *Geophys. Res. L* **1996**, *23*, 2227–2230. [CrossRef]
28. Tyler, S.C.; Ajie, H.O.; Rice, A.L.; Cicerone, R.J.; Tuazon, E.C. Experimentally determined kinetic isotope effects in the reaction CH_4 with Cl: Implications for atmospheric CH_4. *Geophys. Res. Lett.* **2000**, *27*, 1715–1718. [CrossRef]
29. D'Amico, F.; Ammoscato, I.; Gullì, D.; Avolio, E.; Lo Feudo, T.; De Pino, M.; Cristofanelli, P.; Malacaria, L.; Parise, D.; Sinopoli et al. Integrated analysis of methane cycles and trends at the WMO/GAW station of Lamezia Terme (Calabria, Southern Ita *Atmosphere* **2024**, *15*, 946. [CrossRef]
30. Gorham, K.A.; Abernethy, S.; Jones, T.R.; Hess, P.; Mahowald, N.M.; Meidan, D.; Johnson, M.S.; van Herpen, M.M.J.W.; Xu, Saiz-Lopez, A.; et al. Opinion: A research roadmap for exploring atmospheric methane removal via iron salt aerosol. *Atm Chem. Phys.* **2024**, *24*, 5659–5670. [CrossRef]
31. Buono, A.; Zaccardo, I.; D'Amico, F.; Lapenna, E.; Cardellicchio, F.; Laurita, T.; Amodio, D.; Colangelo, C.; Di Fiore, G.; Giunta, et al. Expanding Continuous Carbon Isotope Measurements of CO_2 and CH_4 in the Italian ICOS Atmospheric Consortium: F Results from the Continental POT Station in Potenza (Basilicata). *Atmosphere* **2025**, *16*, 951. [CrossRef]
32. Heiskanen, J.; Brümmer, C.; Buchmann, N.; Calfapietra, C.; Chen, H.; Gielen, B.; Gkritzalis, T.; Hammer, S.; Hartman, S.; Her M.; et al. The Integrated Carbon Observation System in Europe. *Bull. Am. Meterol. Soc.* **2022**, *103*, E855–E872. [CrossRef]
33. Lapenna, E.; Buono, A.; Mauceri, A.; Zaccardo, I.; Cardellicchio, F.; D'Amico, F.; Laurita, T.; Amodio, D.; Colangelo, C.; Di Fi G.; et al. ICOS Potenza (Italy) Atmospheric Station: A New Spot for the Observation of Greenhouse Gases in the Mediterran Basin. *Atmosphere* **2025**, *16*, 57. [CrossRef]

 https://doi.org/10.3390/environments13030

Laurita, T.; Mauceri, A.; Cardellicchio, F.; Lapenna, E.; De Rosa, B.; Trippetta, S.; Mytilinaios, M.; Amodio, D.; Giunta, A.; Ripepi, E.; et al. CIAO observatory main upgrade: Building up an ACTRIS compliant aerosol in-situ laboratory. *Atmos. Meas. Tech.* **2025**, *18*, 2373–2396. [CrossRef]

Madonna, F.; Amodeo, A.; Boselli, A.; Cornacchia, C.; Cuomo, V.; D'Amico, G.; Giunta, A.; Mona, L.; Pappalardo, G. CIAO: The CNR-IMAA advanced observatory for atmospheric research. *Atmos. Meas. Tech.* **2011**, *4*, 1191–1208. [CrossRef]

Villalobos, Y.; Gomez-Ortiz, C.; Scholze, M.; Monteil, G.; Karstens, U.; Fiore, A.; Brunner, D.; Thanwerdas, J.; Cristofanelli, P. Towards improving top-down national CO_2 estimation in Europe: Potential from expanding the ICOS atmospheric network in Italy. *Environ. Res. Lett.* **2025**, *20*, 054002. [CrossRef]

Keeling, C.D. The concentration and isotopic abundances of atmospheric carbon dioxide in rural areas. *Geochim. Cosmochim. Acta* **1958**, *13*, 322–334. [CrossRef]

Mona, L.; Amodeo, A.; Pandolfi, M.; Pappalardo, G. Saharan dust intrusions in the Mediterranean area: Three years of Raman lidar measurements. *J. Geophys. Res. Atmos.* **2006**, *111*, D16203. [CrossRef]

De Rosa, B.; Mytilinaios, M.; Amodeo, A.; Colangelo, C.; D'Amico, G.; Dema, C.; Gandolfi, I.; Giunta, A.; Gumà-Claramunt, P.; Laurita, T.; et al. Observations of Saharan Dust Intrusions over Potenza, Southern Italy, During 13 Years of Lidar Measurements: Seasonal Variability of Optical Properties and Radiative Impact. *Remote Sens.* **2025**, *17*, 453. [CrossRef]

Andrews, A.E.; Kofler, J.D.; Trudeau, M.E.; Williams, J.C.; Neff, D.H.; Masarie, K.A.; Chao, D.Y.; Kitzis, D.R.; Novelli, P.C.; Zhao, C.L.; et al. CO_2, CO and CH_4 measurements from the NOAA Earth System Research Laboratory's Tall Tower Greenhouse Gas Observing Network: Instrumentation, uncertainty analysis and recommendations for future high-accuracy greenhouse gas monitoring efforts. *Atmos. Meas. Tech.* **2014**, *7*, 647–687. [CrossRef]

Zhu, X.; Di, D.; Ma, M.; Shi, W. Stable isotopes in greenhouse gases from soil: A review of theory and application. *Atmosphere* **2019**, *10*, 377. [CrossRef]

Craig, H. Isotopic standards for carbon and oxygen and correction factors for mass-spectrometric analysis of carbon dioxide. *Geochim. Cosmochim. Acta* **1957**, *12*, 133–149. [CrossRef]

Dunn, P.J.H.; Malinovsky, D.; Ogrinc, N.; Potočnik, D.; Flierl, L.; Rienitz, O.; Paul, D.; Meijer, H.A.J. Re-determination of $R(^{13}C/^{12}C)$ for Vienna Peedee belemnite (VPDB). *Rapid Commun. Mass Spectrom.* **2024**, *38*, e9773. [CrossRef]

Chen, H.; Winderlich, J.; Gerbig, C.; Hoefer, A.; Rella, C.W.; Crosson, E.R.; Van Pelt, A.D.; Steinbach, J.; Kolle, O.; Beck, V.; et al. High-accuracy continuous airborne measurements of greenhouse gases (CO_2 and CH_4) using cavity ring-down spectroscopy (CRDS) technique. *Atmos. Meas. Tech.* **2020**, *3*, 375–386. [CrossRef]

Buono, A.; Zaccardo, I.; D'Amico, F.; Mona, L.; Ripepi, E.; Dema, C. Continuous $\delta^{13}C$-CO_2 and $\delta^{13}C$-CH_4 Measurements in the Atmosphere—Data Gathered at Potenza (Basilicata, Southern Italy) in 2025. Version 1. Available online: https://hub.itineris.cnr.it/datasets/ (accessed on 15 December 2025).

Altman, D.G.; Bland, J.M. Measurement in Medicine: The Analysis of Method Comparison Studies. *J. R. Stat. Soc. Ser. D* **1983**, *32*, 307–317. [CrossRef]

Bland, J.M.; Altman, D.G. Statistical methods for assessing agreement between two methods of clinical measurement. *Lancet* **1986**, *327*, 307–310. [CrossRef]

Datta, D. *Blandr: Bland-Altman Method Comparison Package for R, Version 0.6.0*; Zenodo: Geneva, Switzerland, 2024. [CrossRef]

EUMETSAT Viewer. Available online: https://view.eumetsat.int/productviewer?v=mumi:wideareacoverage_rgb_dust (accessed on 20 November 2025).

Lensky, I.M.; Rosenfeld, D. Clouds–Aerosols–Precipitation Satellite Analysis Tool (CAPSAT). *Atmos. Chem. Phys.* **2008**, *8*, 6739–6753. [CrossRef]

Laj, P.; Lund Myhre, C.; Riffault, V.; Amiridis, V.; Fuchs, H.; Eleftheriadis, K.; Petäjä, T.; Salameh, T.; Kivekäs, N.; Juurola, E.; et al. Aerosol, Clouds and Trace Gases Research Infrastructure (ACTRIS): The European Research Infrastructure Supporting Atmospheric Science. *Bull. Am. Meterol. Soc.* **2024**, *105*, E1098–E1136. [CrossRef]

Stein, A.F.; Draxler, R.R.; Rolph, G.D.; Stunder, B.J.B.; Cohen, M.D.; Ngan, F. NOAA's HYSPLIT atmospheric transport and dispersion modeling system. *Bull. Am. Meteorol. Soc.* **2015**, *96*, 2059–2077. [CrossRef]

National Oceanic and Atmospheric Administration Air Resources Laboratory. Real-Time Environmental Applications and Display sYstem. Available online: https://www.ready.noaa.gov (accessed on 23 January 2026).

Müller, T.; Laborde, M.; Kassell, G.; Wiedensohler, A. Design and Performance of a Three-Wavelength LED-Based Total Scatter and Backscatter Integrating Nephelometer. *Atmos. Meas. Tech.* **2011**, *4*, 1291–1303. [CrossRef]

ACTRIS Cloudnet Data Portal. Available online: https://cloudnet.fmi.fi (accessed on 10 December 2025).

ACTRIS Earlinet Data Portal. Available online: https://data.earlinet.org/earlinet (accessed on 26 January 2026).

Farquhar, G.D.; Ehleringer, J.R.; Hubick, K.T. Carbon Isotope Discrimination and Photosynthesis. *Annu. Rev. Plant Physiol. Plant Mol. Biol.* **1989**, *40*, 503–537. [CrossRef]

https://doi.org/10.3390/environments13030145

58. González-Toril, E.; Osuna, S.; Viúdez-Moreiras, D.; Navarro-Cid, I.; Díaz Del Toro, S.; Sor, S.; Bardera, R.; Puente-Sánchez, F.; Diego-Castilla, G.; Aguilera, Á. Impacts of Saharan Dust Intrusions on Bacterial Communities of the Low Troposphere. *Sci.* **2020**, *10*, 6837. [CrossRef]
59. Navarro, A.; del Moral, A.; de Pablos, I.; Delgado, R.; Párraga, J.; Martín-García, J.M.; Martínez-Checa, F. Microorganisms Isola from Saharan Dust Intrusions in the Canary Islands and Processes of Mineral Atmospherogenesis. *Appl. Sci.* **2024**, *14*, 1 [CrossRef]
60. Pataki, D.E.; Ehleringer, J.R.; Flanagan, L.B.; Yakir, D.; Bowling, D.R.; Still, C.J.; Buchmann, N.; Kaplan, J.O.; Berry, J.A. T application and interpretation of Keeling plots in terrestrial carbon cycle research. *Glob. Biogeochem. Cycles* **2003**, *17*, 1 [CrossRef]
61. Bonasoni, P.; Cristofanelli, P.; Calzolari, F.; Bonafe, U.F.; Evangelisti, F.; Stohl, A.; Sajani, S.Z.; van Dingenen, R.; Colombo Balkanski, Y. Aerosol-ozone correlations during dust transport episodes. *Atmos. Chem. Phys.* **2004**, *4*, 1201–1215. [CrossRef]

 https://doi.org/10.3390/environments1303C

MDPI AG
Grosspeteranlage 5
4052 Basel
Switzerland
Tel.: +41 61 683 77 34

MDPI Books Editorial Office
E-mail: books@mdpi.com
www.mdpi.com/books

www.ingramcontent.com/pod-product-compliance
Lightning Source LLC
Chambersburg PA
CBHW041953260926
51850CB00019B/2706
* 9 7 8 3 7 2 5 8 7 7 4 7 8 *